Innovative
Management
in the DARPA High Altitude Endurance
Unmanned Aerial Vehicle Program

PHASE II EXPERIENCE

Jeffrey A. Drezner

Geoffrey Sommer

Robert S. Leonard

T0195527

DEFENSE ADVANCED RESEARCH PROJECTS AGENCY

RAND

NATIONAL DEFENSE RESEARCH INSTITUTE

The research described in this report was sponsored by the Defense Advanced Research Projects Agency. The research was conducted in RAND's National Defense Research Institute, a federally funded research and development center supported by the Office of the Secretary of Defense, the Joint Staff, the unified commands, and the defense agencies under Contract DASW01-95-C-0059.

Library of Congress Cataloging-in-Publication Data

Drezner, Jeffrey A.
   Innovative management in the DARPA high-altitude endurance
unmanned aerial vehicle program : phase II experience / Jeffrey A.
Drezner, Geoffrey Sommer, Robert S. Leonard.
       p.   cm.
   "Prepared for the Defense Advanced Research Project Agency
(DARPA) by RAND's National Security Research Division."
   "MR-1054-DARPA."
   ISBN 0-8330-2717-4
   1. Drone aircraft. I. Sommer, Geoffrey, 1957- .
II. Leonard, Robert S. III. Title.
UG1242.D7D74   1999
623.7 ' 469—dc21   99-19844
                                 CIP

Cover photo image of Global Hawk courtesy of Teledyne Ryan Aeronautical.

Published 1999 by RAND
1700 Main Street, P.O. Box 2138, Santa Monica, CA 90407-2138
1333 H St., N.W., Washington, D.C. 20005-4707
RAND URL: http://www.rand.org/
To order RAND documents or to obtain additional information,
contact Distribution Services: Telephone: (310) 451-7002;
Fax: (310) 451-6915; Internet: order@rand.org

The High Altitude Endurance Unmanned Aerial Vehicle (HAE UAV) program, a joint program conducted under the direction of the Defense Advanced Research Projects Agency (DARPA), incorporates a number of innovative elements in its acquisition strategy. The objectives of this research are to understand how the various innovations affect the program outcomes and to identify the lessons of the HAE UAV program that might be applied to a wider variety of projects to improve DoD acquisition strategies.

The HAE UAV program includes two air vehicle programs, the Tier II Plus Global Hawk and the Tier III Minus DarkStar, and a Common Ground Segment. The program is divided into four phases. This study was initiated in 1994. The RAND study approach is to observe and report on the program, phase by phase. A report covering the Phase I experience of the Global Hawk was issued in 1997.[1] This report covers the Phase II experience of all three components of the HAE UAV program; neither DarkStar nor the Common Ground Segment experienced Phase I. The information is complete through August 1998.

The reports covering the HAE UAV program are intended to be cumulative; that is, each successive report provides coverage of the entire program up to that point. Thus, some of the data presented in the earlier report on Phase I of the Global Hawk program is reproduced here in an abbreviated form.

---

[1]Sommer et al., *The Global Hawk Unmanned Aerial Vehicle Acquisition Process: A Summary of Phase I Experience,* Santa Monica, Calif.: RAND, MR-809-DARPA, 1997.

This report should be of interest to analysts and decisionmakers concerned with reforming the defense acquisition process.

This research was sponsored by the HAE UAV Joint Program Office in DARPA and was conducted in the Acquisition and Technology Policy Center of RAND's National Defense Research Institute, a federally funded research and development center (FFRDC) sponsored by the Office of the Secretary of Defense, the Joint Staff, the unified commands, and the defense agencies.

CONTENTS

TABLES

Improving acquisition policy, processes, and management requires the accumulation of experience from ongoing or recently completed projects, especially those involving unusual situations or innovative acquisition policies. This research contributes to that understanding through its close work with the DARPA High Altitude Endurance (HAE) Unmanned Aerial Vehicle (UAV) program office, whose acquisition strategy represents a radical departure from normal Department of Defense (DoD) procedures. The objectives of this research are to understand how the innovations affect program outcomes and to identify the lessons of the HAE UAV program that might be applied to a wider variety of projects to improve DoD acquisition strategies.

The United States has a poor development track record with UAVs. Technical problems have led to cost growth and schedule slip, as well as disappointing operational results. Examples include the Army's Lockheed Aquila UAV, which was canceled in the late 1980s, and the Teledyne Ryan Medium Range UAV (BQM-145A), which was canceled in October 1993. The causes of this poor historical experience are not clear. One element may be lackluster support from potential users because of the lack of an operational constituency in the military services. Another factor is an apparent underestimation of technical risk in UAV programs, perhaps deriving from their "model airplane" heritage.

The HAE UAV program grew out of a Joint Requirements Oversight Council (JROC)-endorsed three-tier approach to satisfying UAV requirements. Tier I was implemented as the short-range *Gnat 750,* and the Tier II medium-altitude endurance requirement was implemented as the *Predator*. The Tier III requirement for long-endurance reconnaissance, surveillance, and target acquisition was left unsatisfied due to cost considerations. The Tier II+/Tier III– HAE UAV program was initiated as a more cost-effective way to meet the requirement. Tier II+ is the conventional UAV known as Global Hawk; Tier III– is the DarkStar low-observable UAV. A third element of the program, added after the program began, is the Common Ground Segment (CGS).

The HAE UAV program was structured to address problems that had plagued past UAV development efforts. The program's innovative acquisition strategy featured several key elements:

- The program was designated as an Advanced Concept Technology Demonstration (ACTD). This allowed use of a streamlined management and oversight process, provided for early participation of the user community, and bound the schedule length. The goal of the ACTD was to demonstrate military utility in a relatively short timeframe. The use of mature technology was intended to limit risk.

- Section 845 Other Transaction Authority (OTA) allowed DARPA to waive almost all traditional acquisition rules and regulations. The result was a tailored program structure with increased contractor design responsibility and management authority.

- Integrated Product and Process Development (IPPD) used Integrated Product Teams (IPTs) to manage the program. Each IPT included contractor and government personnel who worked together to resolve issues.

- Cost was the single firm requirement. The HAE UAV program had only one requirement: a unit flyaway price (UFP) of $10 million for air vehicles 11–20 (in FY94 dollars). All other performance characteristics were stated as goals and could be traded to achieve the UFP target.

This combination of innovations was unique to the HAE UAV program. While other programs included ACTD or IPPD experience, the HAE UAV was the only program to combine all of these elements into an acquisition strategy.

OUTCOMES TO DATE

The HAE UAV program began in June 1994 with the issuance of a sole-source award to Lockheed Martin Skunk Works (LMSW) for the Tier III– DarkStar and release of a draft Request for Proposal (RFP) for a Phase I competition for Tier II+. Phase II for the Tier II+ began in June 1995 with the selection of the Teledyne Ryan Aeronautical (TRA) Global Hawk. At the time of this writing, the systems were approaching the end of Phase II, which included development and test of two air vehicles of each system. The Air Force will assume program management from DARPA at the beginning of FY99, which also marks the beginning of the Phase III user evaluation. The ACTD program is planned to end in December 1999 with a force-mix and military-utility decision.[2]

As of August 1998, program outcomes have been mixed. The total ACTD program budget of just under $1 billion has been maintained, but developmental costs grew substantially in all three segments of the program. Development costs to the government for the original Phase II statement of work, after accounting for contractor cost share, have increased by 44 percent, 44 percent, and 32 percent for Global Hawk, DarkStar, and CGS respectively. Total program costs have increased even more as additional program activities have been added. As a result, some planned developmental activities have been eliminated from the program, while others have been added to address the technical problems encountered. To maintain the overall budget, fewer assets will be available to support the user demonstration and evaluation. The number of Global Hawks fabricated during the ACTD program has been reduced from ten to five, the number of DarkStars from ten to four, and the number of ground segments from three to two.

The overall schedule has been maintained, but the breadth and depth of information available to support the force-mix and military-utility decisions have been reduced. The ACTD program is still planned to end in December 1999. However, substantial slips in in-

---

[2]The cost and schedule information presented in this report, as well as management information and planning, were current as of August 1998. We recognize that changes have occurred between the analytical cutoff date and the report's publication. We will update information in our subsequent Phase III report.

terim milestones, including the first flight of Global Hawk, the resumption of flight tests for DarkStar, and completion of Phase II activities have significantly reduced the Phase III user demonstration and evaluation period. Phase III was originally planned to be 24 months in length. Currently, Phase III is planned to be 15 months long (October 1998–December 1999), with the first several months dedicated to completion of Phase II test activities. User demonstration tests have effectively been reduced to 12 months (January–December 1999) and now include only a limited range of conditions and scenarios. The abbreviated development test and user evaluation threatens to reduce significantly the amount and type of information available to support the force-mix and military-utility decisions.

The performance of all three program segments will likely be close to the original goals. However, the UFP requirement will be breached for both Global Hawk and DarkStar. An ambiguous definition of military utility has resulted in a reluctance on the part of both the government and the contractors to drop functionality to maintain UFP. Because military utility was undefined, the contractors had little guidance on what functionality could be dropped without threatening utility.

Several events and conditions precipitated these outcomes. The program may not meet the criteria established in ACTD policy concerning use of existing mature technology. Both air vehicles are new designs that include capabilities not previously demonstrated. Similarly, the software-development and system-integration challenges were underestimated, particularly regarding the difficulty of integrating commercial or off-the-shelf components and subsystems.

The crash of the first DarkStar on takeoff during its second flight in April 1996 was an important event that affected all three elements of the HAE UAV program. Characteristics unique to the acquisition strategy lay beneath the technical reasons for the crash. The already-accelerated development schedule was pushed even harder by the contractor (LMSW). One result was less-than-adequate aerodynamic information regarding the DarkStar configuration. Another contributing factor was poor judgment regarding risks, even after a discrepancy was found between data from the first flight and model predictions. The value of prior experience was overestimated. The crash resulted in increased risk aversion for all program segments, increased reviews and failure analysis, and greater emphasis on operator training. Costs and schedules grew, particularly in Global Hawk and DarkStar, as the government added activities to ensure adequate development and reduced risk.

Our analysis includes a comparison of the costs and schedules of the Global Hawk and DarkStar with aircraft, cruise missile, and other UAV programs. Such analyses are problematic in that every program is unique, particularly in terms of its development phases. Nevertheless, the costs and schedules of the Global Hawk and DarkStar are about what we would expect, given the activity content of the program. Development costs are comparable to the Light Weight Fighter and Have Blue technology demonstration programs (in inflation adjusted dollars), but much more is being accomplished in system integration and operational suitability. Schedule lengths are approximately the same as for the F-117 and F-16 programs, including time to first flight and projected time to first operational delivery. However, the HAE UAV systems at the end of the ACTD program will be less mature than typical aircraft programs at the end of an engineering and manufacturing development (EMD) phase.

ASSESSMENT OF THE STRATEGY

Assessing the effect of the innovative attributes of the acquisition strategy on program outcomes is a challenging task prohibiting any simple explanation. Each element of the strategy has advantages (benefits) and disadvantages (costs). Complex interactions among the various elements make it difficult to distinguish the effect of one factor over another. These interactions produce both positive and negative effects. Additionally, other elements of the acquisition strategy, unrelated to the innovative attributes, also affect program outcomes. Lastly, the acquisition environment itself is too complex for any strategy to resolve completely.

We believe that the ACTD designation, which bound program length, may have resulted in an accelerated program structure that may have left insufficient time to determine military utility. However, it increased design and decision flexibility, provided an opportunity for (if not the actuality of) early user involvement, and had the potential to demonstrate a new capability faster than traditional acquisition programs.

Section 845 OTA brought lower overhead costs by eliminating complex reporting and auditing processes. Additionally, decisions were made faster because the contractors owned the decision processes. We also believe that the good working environment created in the program and relatively low cost of the systems are the result of the use of Section 845 OTA. On the other hand, the flexibility allowed in systems-engineering processes resulted in inadequate engineering discipline in both air-vehicle segments, resulting in some of the technical problems.

The government-industry relationship improved over the course of the program, characterized by reduced oversight and open interactions. The IPT structure allowed the government to obtain timely insight into program status, problems, and solutions, as well as provide timely input into the contractors' decision processes. However, the IPPD process, which encourages teamwork, potentially conflicts with Section 845 OTA, which encourages contractor responsibility. The issue arises from the participation of joint program office (JPO) representatives on contractor IPTs when the contractor has decision authority. Under these conditions, who owns the process and who is accountable for decisions is not always clear. While this conflict may not be inherent in the two processes (IPPD and Section 845 OTA), it is inherent in the way those processes were implemented. An optimum balance between oversight and technical participation has not yet been achieved.

The UFP requirement, with all other system characteristics stated as performance goals, was a different, unfamiliar approach to system design for the contractors. This approach, combined with other elements of the acquisition strategy, theoretically gave the contractors complete control over the cost-performance tradespace. Because of the ambiguous definition of military utility, both the contractor and the government were reluctant to reduce functionality as a way of maintaining UFP.

A related issue concerns the relationship between UFP and non-recurring engineering (NRE) funds. A certain amount of NRE activity must be performed, and a certain level of investment must be made, to achieve a given UFP. The balance between NRE funding and the UFP must be clearly rationalized for the UFP to be both credible and achievable. To mitigate an apparent NRE funding shortfall, the contractors "created" additional NRE funding by reducing developmental activity (which also minimized developmental cost growth). In contrast, the acquisition approach intended that functionality be traded to meet UFP.

While the UFP is likely to be breached when production HAE UAV aircraft are built, significant UFP growth should not be considered a failure. Even a 50-percent growth in UFP for either the Global Hawk or DarkStar (i.e., a $15-million UFP) could still result in a capable and cost-effective solution to the mission need.

CONCLUSIONS AND LESSONS LEARNED

Problems should be expected in a first-time application of any radical innovation. On balance, both the JPO and the contractors rate the acquisition strategy highly. We also observe that the high degree of process flexibility inherent in this acquisition strategy requires high-quality JPO personnel, sustained senior management support within the relevant DoD organizations, and a cooperative contractor willing to accept increased responsibility. Thus, the strategy may not be applicable to all acquisition programs. However, we believe overall that the strategy is an effective alternative to traditional acquisition processes and could be applied more widely in the acquisition community. We have identified several ways in which the strategy can be enhanced in future applications.

Section 845 OTA provided tangible benefits to both the program office and contractors, including less-burdensome and more-informal management processes, reduction in overhead costs, and an improved work environment. We believe that Section 845 OTA (Section 804 for the military services) also is more widely applicable. The key to successful use of this increased flexibility will be to strike a balance between increased contractor management authority and more traditional oversight mechanisms. In particular, flexible government management, tailored for each contractor, is required to offset contractor weaknesses. While reliance on the contractor for design and management can save time and money, the government must be able to step in when the contractor demonstrates weaknesses in key areas. A mechanism for this type of intervention should be incorporated into future agreements defining the program and government-industry relationships.

We believe that the mechanisms for user participation should be more formalized. Coordination among user groups should be improved and expectations clarified. Military utility can be defined early in the program to provide better guidance to contractors regarding the priority ranking of potential system capabilities.

Thorough planning of the initial program structure, to assess and plan for risks, would improve execution. Under the innovative acquisition strategy used in the HAE UAV program, improved planning and risk management are the responsibility of both the government and the contractors. Many of the problems experienced in the HAE UAV program can in hindsight be attributed to the relatively high risk that was tacitly accepted by the program.

We believe that future programs should evolve beyond the limitation of UFP as the single requirement. Cost, schedule, and performance can all be goals to be traded against each other to achieve an optimal solution for the military mission. Programs can set boundaries for cost, schedule, and performance parameters, but the resulting trade space must be large enough to enable realistic and credible tradeoffs. This flexibility is perhaps appropriate in the near future only for small programs that incorporate new capabilities and concepts, and should be tried several times on an experimental basis before being applied more ambitiously. We believe that this approach can better produce a cost-effective capability in a shorter period of time.

ACKNOWLEDGMENTS

This research required extraordinary access to personnel and documentation from both the DARPA HAE UAV Joint Program Office and the prime contractors for each of the main program components: Lockheed Martin Skunk Works (DarkStar), Teledyne Ryan Aeronautical (Global Hawk), and Raytheon E-Systems (CGS). We greatly appreciate the time, cooperation, and candor of government and industry officials during our many discussions and in reviewing draft reports.

Our RAND colleagues Tim Bonds and Mark Lorell provided helpful reviews of several interim products as well as the final report. Their comments and suggestions are appreciated.

The authors are solely responsible for any remaining errors.

ACTD	Advanced Concept Technology Demonstration
ASC	Aeronautical Systems Command
ATD	Advanced Technology Demonstrations
AUSV	Association of Unmanned Space Vehicles
CAIV	Cost As an Independent Variable
CAP	Contractor-Acquired Property
CGS	Common Ground Segment
CINC	Commander in Chief
CLIN	Contract Line Item Number
CONOPS	Concept of Operations
CONV	Conventional
COTS	Commercial-Off-The-Shelf
CPFF	Cost Plus Fixed Fee
CPIF	Cost Plus Incentive Fee
CR	Close Range
DARO	Defense Airborne Reconnaissance Office
DARPA	Defense Advanced Research Projects Agency

DFARS   Defense Federal Acquisition Regulations
               Supplement

DoD   Department of Defense

DODD   Department of Defense Directive

DPE   DarkStar Processing Element

DSB   Defense Science Board

DUSD (AT)   Deputy Undersecretary of Defense (Advanced
               Technology)

EIC   Element Integration Complete

EMD   Engineering and Manufacturing Development

EO/IR   Electro-Optical/Infra-red

FARs   Federal Acquisition Regulations

FDR   Final Design Review

FF   First Flight

FRR   Flight Readiness Review

FTE   Full Time Equivalent

GFE   Government-Furnished Equipment

HAE UAV   High Altitude Endurance Unmanned Air Vehicle

HALE   High Altitude Long Endurance

IDR   Initial Design Review

IMP   Integrated Master Plan

IMS   Integrated Master Schedule

IPPD   Integrated Product and Process Development

IPT   Integrated Product Team

IRT   Independent Review Team

JPO	Joint Program Office
JROC	Joint Requirements Oversight Council
LADC	Lockheed Advanced Development Company
LCRS	Launch, Control, and Recovery System
LMSW	Lockheed Martin Skunk Works
LO	Low Observable
LOS	Line Of Sight
LRE	Launch and Recovery Element
LRIP	Low Rate Initial Production
MAE	Medium Altitude Endurance
MCE	Mission Control Element
MDAP	Major Defense Acquisition Project
MILSPEC	Military Specification
MNS	Mission Need Statement
MoU	Memorandum of Understanding
NRE	Nonrecurring Engineering
NTE	Not-To-Exceed
O&M	Operations and Maintenance
OSD	Office of the Secretary of Defense
OTA	Other Transactions Authority
OTS	Off-The-Shelf
PDS	Processing and Display System
PLSS	Precision Locator Strike System
PM	Project Manager

PO	Program Office
PRR	Production Readiness Review
QAR	Quality Assurance Report
RFP	Request For Proposal
RSTA	Reconnaissance, Surveillance, and Target Acquisition
SAR	Synthetic Aperture Radar
SCD	System Compatibility Design
SETA	Systems Engineering and Technical Analysis
SIGINT	Signals intelligence
SIL	System Integration Lab
SOW	Statement of Work
TDD	Task Description Document
TINA	Truth in Negotiations Act
TPM	Technical Performance Measures
TRA	Teledyne Ryan Aeronautical
UAV	Unmanned Aerial Vehicle
UFP	Unit Flyaway Price
USACOM	U.S. Atlantic Command
USD (AT)	Undersecretary of Defense (Advanced Technology)
VLO	Very Low Observable
WBS	Work Breakdown Structure
WPAFB	Wright-Patterson Air Force Base

Chapter One

INTRODUCTION

The Defense Advanced Research Projects Agency (DARPA), in conjunction with the Defense Airborne Reconnaissance Office (DARO), is developing two unmanned air vehicles (UAVs) intended to provide surveillance information to the warfighter. They respond to the recommendations of the Defense Science Board (DSB) and to the operational needs stated by DARO on behalf of the military service users. The development strategy used in this program includes several highly innovative attributes intended to reduce cost, accelerate deployment, and engage the user community.

The High Altitude Endurance (HAE) UAV program consists of three complementary system development efforts. The Tier II Plus Global Hawk is a conventionally configured air vehicle. Tier III Minus DarkStar incorporates low-observable technology into the design of the air vehicle. The Common Ground Segment (CGS) provides launch, recovery, mission control, and data processing for both air vehicles. At the time of this writing, the systems were in transition from Phase II, in which development and initial flight testing are performed, to Phase III, which consists of the user demonstration and fabrication of additional air vehicles to support the flight-test program.

BACKGROUND

It is important to place the HAE UAV program in the context of past UAV acquisition experience, because the effects of the unique features of this program may depend to some extent on its being a UAV program. Past experience with UAVs falls into two distinct categories: developmental and operational.

Historically, UAV development programs in the United States have been bedeviled by cost growth, schedule slippage, manifold technical deficiencies during protracted development, and generally disappointing operational results. Notorious examples include the U.S. Army's Lockheed *Aquila* UAV program, which was canceled in the late 1980s,[1] and the Teledyne Ryan BQM-145A Medium Range UAV,[2] which was canceled in October 1993. Both programs suffered from performance and technical problems as well as substantial cost growth.

Overseas, however, operational experience with UAVs has sometimes been good. Israel had excellent results employing *Scout* and *Mastiff* UAVs against Syrian air defenses in the Bekaa Valley in Lebanon during 1982.[3] The Syrians lost 19 surface-to-air missile (SAM) batteries and 86 combat aircraft, essentially a complete air defense system, and the Israeli UAVs played a pivotal role in this accomplishment. Thus, it is generally acknowledged that U.S. UAVs have suffered from flaws in program execution rather than concept.

The cause of the poor developmental track record of UAV programs in the United States is uncertain. The mere fact that they are *unmanned* vehicles cannot be the cause; the United States has had great success with other unmanned systems, including interplanetary spacecraft, satellites, cruise missiles, and submersibles. What, then, makes UAVs unique? One possible explanation is that UAVs in general have never had the operational user support necessary to allow their procurement in sufficient quantities (perhaps because of funding competition from existing programs, or because of the conjectural nature of their capabilities). Thus, the learning curve is never ascended, multiple failures occur, risk tolerance decreases, unit costs rise as a result, and user support decreases yet further in a diminishing spiral.

---

[1]General Accounting Office, *Aquila Remotely Piloted Vehicle—Its Potential Battlefield Contribution Still in Doubt*, GAO/NSIAD-88-19, October 1987.

[2]General Accounting Office, *Unmanned Aerial Vehicles—Medium Range System Components Do Not Fit*, GAO/NSIAD-91-2, March 1991.

[3]Air Chief Marshal Sir Michael Armitage, *Unmanned Aircraft*, London: Brassey's Defense Publishers, 1988, p. 85.

The risk of UAV program failure may also have been compounded by three other factors:

- expectations of low cost (stemming from the "model airplane" heritage of UAVs)

- a flight-safety dichotomy (UAVs need not be "man-rated," but range-safety and redundancy considerations tend to increase costs)

- the variable and unpredictable nature of the aerial environment itself (compared to more predictable oceanic or space environments).

For this study, we draw two cautionary notes from the above. First, one can expect the relative success of UAV programs to be dependent on levels of operational user support. Second, near-term UAV programs may need to be structured in a more risk-averse form than is commonly appreciated. These factors must be considered when weighing the effect of acquisition streamlining and other features applied to the HAE UAV program.

High-level support for UAVs persists, and a strong motivation exists to overcome the programmatic and technical difficulties experienced to date. To overcome these historical problems, DARPA, with congressional support, is conducting an innovative acquisition program that is different from normal Department of Defense (DoD) acquisition procedures in several important ways.

First, the approach gives flexibility to depart from acquisition-specific law and related regulations. Such procedures may, but need not, be employed. Contractors are encouraged to tailor or "reinvent" the acquisition system in ways especially suited to this particular program. The idea is to avoid rigid procurement practices, to encourage use of commercial practices and products, and to encourage firms to organize the project around integrated product teams rather than by functional discipline.

Second, experience indicates that unit price goals have been missed because initial performance expectations were too demanding and constraining. These expectations left little room for design trades in the critical early program phases. In this program, the DARPA/DARO paradigm is to establish a firm cap of $10 million on unit flyaway price (UFP), and let the contractors trade all other performance goals as necessary to stay within that cap. Military capability will be determined through flight test and operational demonstrations, and the program will not transition to the acquisition process if the system does not provide adequate military utility. Note that one result of this program structure is a possible multiyear gap between the end of the program and the beginning of subsequent development or production activity.

Finally, the HAE UAV program has been designated an Advanced Concept Technology Demonstration (ACTD), a program intended to demonstrate mature or maturing technologies to the warfighters in an accelerated fashion. This designation places a premium on early operational user involvement and defines an aggressive program schedule, which in turn should drive (and hopefully limit) program nonrecurring engineering (NRE) costs.

The UAV in the United States Through 1993[4]

In the United States, the first large-scale use of nonlethal UAVs occurred during the Vietnam War, when air-launched Teledyne Ryan UAVs were employed on reconnaissance missions over North Vietnam. Although loss rates were fairly high, useful imagery was recovered, and some UAVs survived for many missions over hostile territory. Of course, the wartime imperative and high production rates resulted in a relatively steep learning curve and consequent reductions in procurement cost.

The Teledyne Ryan family of UAVs (or drones, or Remotely Piloted Vehicles, as they were then known) subsequently expanded to include high-altitude, long-endurance (HALE) variants. A program known as *Compass Arrow* (AQM-91A) was pursued in the 1960s, resulting in an ultra-high-altitude (80,000 ft) surveillance aircraft that was, unfortunately, expensive. The government terminated the *Compass Arrow* program in the early 1970s after 28 aircraft had been

---

[4]This section repeats the discussion in Sommer et al., *The Global Hawk Unmanned Aerial Vehicle Acquisition Progess: A Summary of Phase I Experience,* Santa Monica, Calif.: RAND, MR-809-DARPA, pp. 11–13.

produced but before they became operational, as a result of U.S. rapprochement with China (the primary strategic target).[5] The *Compass Cope* program, which followed *Compass Arrow* in the 1970s, was intended to develop a reconnaissance and signals intelligence (SIGINT) HALE UAV. Boeing and Teledyne Ryan competed in a protracted development and fly-off program, marred by the crash of one of the Boeing aircraft. The program was ultimately terminated, largely because of a weakness in operational requirements justification. The fact that a key potential payload, the Precision Locator Strike System (PLSS), was also slated for installation on the manned U-2 probably contributed to the cancellation.[6] This competition between manned and unmanned airborne reconnaissance options continues.

The 1980s saw a flowering of smaller tactical UAV programs. The successes of the Israelis served as a powerful impetus, and the *Scout* was modified and sold to the United States as the *Pioneer*. It quickly became apparent, however, that there was an operational justification for UAVs with improved capabilities, and several companies offered new creations as a result. One such program was the Lockheed *Aquila*, which became notorious for its management failures and cost overruns.[7]

In FY88, Congress directed the consolidation of DoD nonlethal UAV program management. There was a perception that DoD was pursuing multiple, redundant UAV programs, and an integrated management structure was therefore necessary. The UAV Joint Project Office (UAV JPO) was formed and embarked upon a four-element UAV program in response to several approved Mission Need Statements (MNSs).[8] Among these was one approved by the Joint Requirements Oversight Council (JROC) in January 1990 to establish a "Long Endurance Reconnaissance, Surveillance, and Target Acquisition

---

[5]*A History of Teledyne Ryan Aeronautical, Its Aircraft and UAVs,* Teledyne Ryan corporate brochure, 1992, p. 8; and unpublished RAND research by John F. Schank on cost-estimating relationships for airframes of remotely piloted vehicles.

[6]DMS Incorporated, *The RPV/Drones/Targets Market 1975–1985,* 1975, pp. II-37 to II-40.

[7]General Accounting Office, *Aquila Remotely Piloted Vehicle—Its Potential Battlefield Contribution Still In Doubt,* GAO NSIAD-88-19, October 1987.

[8]DoD, *Unmanned Aerial Vehicles (UAV) Master Plan 1992,* 15 April 1992, pp. 6–8.

(RSTA) Capability." The intent was to provide warfighting commanders in chief (CINCs) with the capability to conduct wide-area, near-real-time RSTA, command and control, SIGINT, electronic warfare, and special-operations missions during peacetime and all levels of war. The CINCs would be able to exercise this capability against defended and denied areas over extended periods of time.[9]

The four core UAV programs were the close-range (CR), short-range (SR), medium-range (MR) and endurance UAVs. The UAV JPO suffered an embarrassing reversal with the cancellation of the core medium-range UAV in October 1993. The remaining programs were reorganized into two groups: the Joint Tactical Program (absorbing the CR and SR) and the Endurance Program.[10]

In July 1993 the JROC endorsed a three-tier approach to acquiring an "endurance" capability:

- Tier I: Quick Reaction Capability

- Tier II: Medium Altitude Endurance

- Tier III: "Full Satisfaction" of the MNS.

Tier I and Tier II were implemented as the *Gnat 750* and *Predator* UAVs. In July 1993, the DSB launched the Deep Target Surveillance/Reconnaissance Alternatives Study to address the Tier III requirement. The study focused on imagery support to military operations but concluded that potential Tier III systems would be either too expensive or unable to satisfy the requirement. At this point, DARO substituted the parallel Tier II+/Tier III– approach for Tier III, meeting that requirement with a high/low force mix of complementary systems. Tier II+ and Tier III– were also known as the CONV (conventional) HAE UAV and LO (low observable) HAE UAV systems, respectively.[11] The evolution of these programs is depicted in Figure 1.1. In this report we are concerned with the two elements of the

---

[9]"High Altitude Endurance Unmanned Aerial Vehicle Systems—Program Briefing for Joint Requirements Oversight Council," 8 November 1994.

[10]DoD, *Unmanned Aerial Vehicles 1994 Master Plan*, 31 May 1994, p. 3–29.

[11]"High Altitude Endurance Unmanned Aerial Vehicle Systems—Program Briefing for Joint Requirements Oversight Council," 8 November 1994.

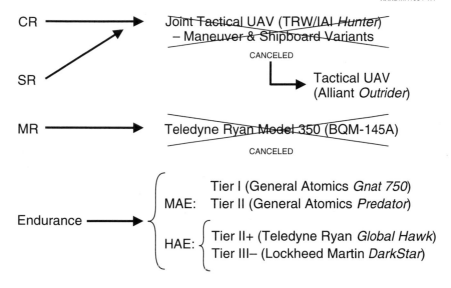

Figure 1.1—Recent UAV Program History

HAE program; CONV and LO, now known as the *Global Hawk* and *DarkStar*.

OBJECTIVES AND APPROACH

Improving acquisition policy, processes, and management requires the accumulation of experience from ongoing or recently completed projects, especially those involving unusual situations or innovative acquisition policies. This research contributes to that understanding through its close work with the HAE UAV program office, whose acquisition strategy represents a radical departure from normal DoD procedures. Our objective has been to understand how the various innovations affected program outcomes and to identify lessons that might be applied to a variety of projects to improve DoD acquisition strategies—not to perform a program assessment.

We have four objectives:

- Observe the HAE UAV program, with particular attention to the execution of the various innovative acquisition strategies being employed.

- Document the program. This provides a useful history for our research, as well as for others.

- Attempt to relate program outcomes, both quantitative (cost, schedule) and qualitative (government-industry relationships, adequacy of oversight), to distinct elements of the programs' acquisition strategies.

- Identify lessons from these programs that are applicable to a variety of programs to improve DoD acquisition processes.

We characterize the HAE UAV program by two sets of factors—the acquisition strategy and "other" factors. The acquisition strategy includes those innovations discussed above, as well as related elements (e.g., a small JPO). Other factors include the quality, culture, and experience level of the contractor(s), degree of technical challenge, funding stability, and adequate identification and management of program risk areas.

The HAE UAV program had several outcomes, measured in both the traditional (cost, schedule, performance) and nontraditional (ease of decisionmaking, use of commercial-off-the-shelf (COTS) technology, quality of government-industry interactions) metrics. Outcomes can be complex, and can be positive and negative. The primary research question is: To what extent can these outcomes, both positive and negative, be explained by the innovative acquisition strategy, as opposed to other factors?

Our research approach was divided into three tasks.

Task 1: HAE UAV Program Tracking. The primary research task was to track and document the experience of both the JPO and contractors as the program proceeded. This task involved periodic discussions with both the JPO and contractors to understand current program status, key events, and milestones, as well as how the innovative elements of the acquisition strategy were being implemented. Through these discussions, we were able to assess whether the acquisition strategy was having the expected effect, as well as identify issues arising in the course of program execution that either affected or were affected by the acquisition strategy. This task also involved a thorough review of program documentation, including solicitations, proposals, agreements, memoranda, and program-review briefings. We also obtained information on program funding and schedules. Our goal was to determine the underlying causes of each major program event, especially events that represented changes from the original plan and expectations.

During FY95, Tier II+ contractors involved in Phase I were interviewed, and additional information was collected from the Phase II contractor regarding transition to that phase. During FY97, we conducted interviews with the prime contractors for each of the three HAE UAV program components:

- Teledyne Ryan Aeronautical (Global Hawk)

- Lockheed Martin Skunk Works (DarkStar)

- Raytheon E-Systems (CGS).

Several discussions with JPO personnel took place during FY97 and FY98 as well.

Task 2: Comparisons with Other Programs. In this portion of the research, we collected and analyzed historical cost, schedule, and performance data from comparable past programs. Relatively little historical data has been preserved on past UAV programs at a detailed level, limiting their value as a baseline for comparison with the current programs. Therefore, we assembled data on program outcomes from broader databases of experience to assess HAE UAV program outcomes in historical context.

Task 3: Analysis and Lessons Learned. As the HAE UAV program proceeds, we will draw together the information collected under Tasks 1 and 2 and present two kinds of overall results. One is focused on understanding the extent to which the HAE UAV program was implemented as planned and the degree to which the program achieved its expected outcomes. The other will be focused on comparisons between the HAE UAV program and other programs. Together, these will provide an understanding of the strengths and weaknesses of the overall HAE UAV acquisition strategy. We will then interpret those results in terms of lessons that might be applied to future programs.

We faced several challenges in this research. First, there is the issue of establishing a baseline for comparison of outcomes and events. We cannot compare the program to what we ideally would want: the same program executed under the traditional acquisition process. Second, case studies permit only limited generalization. In acquisition, each case has unique and hard-to-define characteristics that affect outcomes. Additionally, case studies do not allow control over external influences, though these can often be identified. Lastly, defining "success" is particularly troublesome; the program could succeed in demonstrating the viability of alternative acquisition strategies but still fail to produce the originally planned system. There is also an important difference between deviating from the plan (cost, schedule, and performance outcomes) and doing better than the traditional process.

Given these challenges, we attempt to explain program outcomes based on internal program analyses as well as comparisons with other programs that share salient features. We address these analytic challenges by adhering to a rigorous case-study research design. Our hypothesis is that the innovative attributes of the HAE UAV acquisition strategy significantly affected program outcomes. We followed a data-collection protocol for both industry and JPO discussions. The formal analytic method is a combination of nonequivalent dependent-variable pattern-matching (use of more than one dependent variable), rival-explanation pattern-matching (generation of alternative explanations for an outcome or event), and iterative explanation-building (continuous revision of our conceptual models as we learn more about the program and interact with participants).[12]

We must mention one additional aspect of the HAE UAV program. A feature of the acquisition process used here was the reduction of program documentation, and the participants fully exercised that opportunity. Consequently, virtually no systematic documentation of events exists that can be used to construct cause-effect relationships that can be clearly linked to final outcomes. Instead, we base

---

[12]See Yin, Robert K., *Case Study Research: Design and Methods*, 2nd ed., Thousand Oaks, Calif.: SAGE Publications, 1994.

the following description and analysis largely on discussions with industry and government program managers, drawing on their memory and interpretations.

HAE UAV PROGRAM OVERVIEW AND OUTCOMES

The origins of the detailed objectives and structure of the HAE UAV program are not well documented; the process was not performed under the rules of Department of Defense Directive (DODD) 5000.1, and none of the formal documentation of requirements validation and concept formulation was prepared.[13] However, the broad outline of the program took shape during early 1994. The HAE UAV Program Office was formed in February 1994 and charged with developing a "family of reconnaissance vehicles" in response to the JROC MNS on Long-Endurance RSTA Capability (JROCM-003-90, 9 April 1990). The Defense Airborne Reconnaissance Office (DARO) sponsored the program, and designated DARPA as the executive agent for the initial phases of the ACTD. The Air Force was designated the lead for the final phase on the ACTD program.

The Memorandum of Understanding (MoU) establishing the program made it clear that the program would focus on management issues as well as the technical development of the two systems and associated ground segments.[14] The MoU also specified that the program would be managed in a joint program office (JPO), with a DARPA Program Director and Air Force and Navy Deputy Program Directors. Similarly, the initial version of the HAE UAV ACTD Management Plan, which is the single guiding document for program management, explicitly stated that testing the viability of the acquisition strategy is one of two program objectives.[15] Sub-

---

[13]We do not claim that the early phases of program formulation were not conducted with appropriate rigor or completeness, only that the results are not documented in typically voluminous form.

[14]Specific management issues mentioned include user involvement, affordability, and streamlined development to be accomplished through innovative contracting, a small program office, and informal processes. *Memorandum of Understanding for the High Altitude Endurance Unmanned Aerial Vehicle Among Department of the Army, Department of the Navy, Department of the Air Force, Advanced Projects Research Agency, Defense Airborne Reconnaissance Office,* 12 October 1994.

[15]*HAE UAV ACTD Management Plan, Version 1.0 (draft),* 15 December 1994.

sequent versions of the management plan left this intact while adding objectives.

The U.S. Atlantic Command (USACOM) was identified as the user organization and would define military utility prior to the beginning of the Phase III demonstration. USACOM was also to be a participant in major program reviews and a partner in developing the Concept of Operations (CONOPS).

The program was composed of two complementary air vehicles. The Tier II+ would be a conventional configuration and simultaneously carry both a Synthetic Aperture Radar (SAR) and Electro-Optical/Infrared (EO/IR). The Tier III– would be a low observable configuration and carry either an SAR or EO/IR. The Tier III– would trade some endurance and range for low observability. The performance objectives of the two air vehicles as defined in the initial HAE UAV ACTD management plan is contained in Table 1.1. According to the FY97 UAV Annual Report published by DARO, both the Tier II+ and Tier III– were "envisioned from the start as needing significant development work. . . ." The potential benefits of the systems were judged to be worth the resulting increased risk.

Table 1.1

HAE UAV Performance Objectives

Characteristics	Conventional HAE UAV	Low observable HAE UAV
On-station loiter (hours)	24	>8
Operating radius (n miles)	2000–3000	>500
Loiter altitude (ft msl)	60,000–65,000	>45,000
True air speed (knots)	300–375	>250
Takeoff weight (lb.)	15,000–27,000	8500
Survivability measures	Threat warning, ECM, decoys	Very low observable
Sensor payload	SAR, GMTI and EO/IR	SAR or EO
Sensor payload wt. (lb.)	1000–1500	1000
Command and control	UHF FLEETSATCOM	UHF FLEETSATCOM
Ground control	Max use of GOTS/COTS	Common with Tier II+
Data exploitation	CIGSS, JSIPS/JSIPS-N, CARS, MIES, JICS, and NPIC	Common with Tier II+

SOURCE: HAE UAV ACTD Management Plan, Version 1.0, December 1994 (draft), Table 1.

We should note that the Tier II+ and Tier III– programs were originally planned as separate, stand-alone programs. Pentagon studies
during 1993 supported an air vehicle with performance characteristics like those specified for Tier II+. Lockheed Martin Skunk Works
(LMSW) provided an unsolicited proposal for Tier III– after Tier II+
was approved to proceed. Thus, the two development efforts were
clearly distinct at the outset, though they were managed within the
same framework.

Table 1.2 shows the expected performance characteristics for the
HAE UAV air vehicles toward the end of Phase II. The Global Hawk
and DarkStar are expected to achieve the performance objectives set
at the beginning of the program. The changes involve more precision in the values for some performance parameters, reflecting the
final design configurations. The changes also show that key parameters are expected to be at the high end of the originally estimated
range: payload weight, takeoff weight, operating radius, and loiter
altitude. Neither table includes supportability or reliability parameters; such issues were not a priority. While the performance
objectives are likely to be met, the single requirement—a $10-million
UFP (FY94$) for air vehicles 11–20—is likely to be breached. Cost
performance trades were not made to the extent envisioned because

Table 1.2

HAE UAV Performance Characteristics

Characteristics	Global Hawk	DarkStar
On-station loiter (hours)	24	8
Operating radius (n miles)	3000	500
Loiter altitude (ft)	>60,000	50,000
True air speed (knots)	300–350	300
Takeoff weight (lb.)	25,600	8600
Survivability measures	Threat warning, ECM, decoys	Very low observable
Sensor payload	SAR, GMTI and EO/IR	SAR or EO
Sensor payload wt. (lb.)	1800	1000
Command and control	UHF FLEETSATCOM	UHF FLEETSATCOM
Ground control	Common	
Data exploitation	CIGSS, JSIPS/JSIPS-N, CARS, MIES, JICS, and NPIC	Common with Tier II+

SOURCE: *HAE UAV ACTD Management Plan, Version 7.0*, December 1997 (draft),
Table 1; HAE UAV public-release characteristics table, www.darpa.mil/haeuav/
charac.htm.

of a reluctance, by both the government and the contractors, to drop functionality; the definition of military utility was ambiguous and provided little guidance to the contractors regarding what capabilities the user valued.

Originally, the two program elements, Global Hawk and DarkStar, were to be developed in parallel, although with somewhat different phasing. A Phase I design study and source selection was planned for Global Hawk, but that phase was skipped for DarkStar on the presumption that such work was performed in previous efforts. Each program consisted of a Phase II in which detailed design and engineering development would be conducted, culminating in a test period. Test activities in Phase II were to be independent due to differences in contracting and development schedules. The program was planned to transition to Air Force management at the beginning of Phase III. In Phase III additional units would be produced to support a two-year user evaluation and field demonstration. Presuming that one or both systems proved satisfactory, that would be followed by Phase IV, production for operations.

In Phase II, the contractors would build one ground segment and two of each type of air vehicle. These first air vehicles were called engineering development models. The purpose of the 12-month Phase II flight test was to mature the system and correct mission-critical deficiencies. In Phase III, up to two additional ground segments and eight air vehicles of each type would be built, funding permitting.[16] Thus, residual assets at the end of the ACTD were envisioned as a maximum of three ground segments and 20 air vehicles, all with complete payloads. The Phase-III Agreements for both Tier II+ and Tier III– were to include an "irrevocable offer" to produce units 11–20 in Phase IV for $100 million (FY94$), for an average UFP of $10 million. The initial ACTD management plan stated that competition existed between the two air vehicles for the Phase IV force-mix decision.

The planned schedule, as it existed in mid-1994, is shown in Figure 1.2. Both systems were to participate in a joint-user field demon-

---

[16]*HAE UAV ACTD Management Plan, Version 1.0 (draft)*, 15 December 1994, p.12.

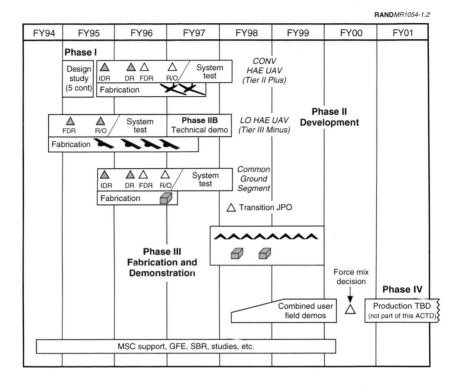

RANDMR1054-1.2

Figure 1.2—HAE UAV Schedule As of Early 1994

stration during 1998 and 1999, leading to a decision on production for operational use at the beginning of 2000.

The detailed phasing of both programs evolved during late 1994 and early 1995, and the notion of a common ground segment that could serve both flight systems became an established part of the program. Thus, by the time Global Hawk was under contract for Phase II (June 1995), the planned schedule was as shown in Figure 1.3. This figure serves as a baseline reference point when examining how the programs actually evolved.

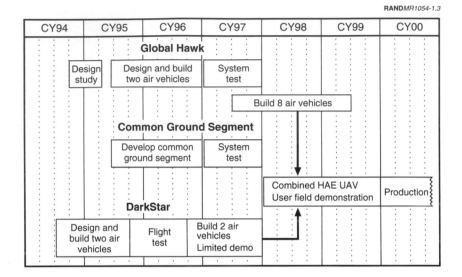

RAND*MR1054-1.3*

Figure 1.3—HAE UAV Program Schedule As of Mid-1995

Figure 1.4 shows a more recent program schedule.[17] By mid-1998, technical problems leading to cost and schedule growth had significantly lengthened Phase II. However, the ACTD program was still planned for completion in December 1999, resulting in a shortening of the Phase III user demonstration and evaluation. The schedule in Figure 1.4 shows Phase III continuing through September 2000, but, as discussed below, the program is unfunded past December 1999. The ACTD program is currently planned to end in December 1999 with military-utility and force-mix decisions.

The transferral of program management from DARPA to Air-Force lead was originally planned for the third quarter of 1997, marking the end of Phase II and the beginning of Phase III user evaluation.[18] The transition is currently planned for September 1998 (a slip of approximately 12 months from the original plan). However, develop-

---

[17]Taken from Heber, "HAE UAV POM 00 Brief," 5 March 1998.

[18]*HAE UAV ACTD Management Plan, Version 1.0 (draft)*, 15 December 1994. The subsequent Phase II Agreement defined the transition point as December 1997, the completion of Phase II activities.

RAND*MR1054-1.4*

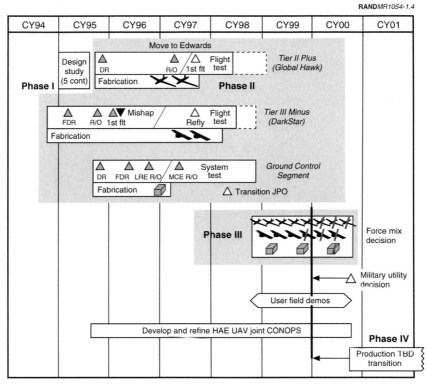

Figure 1.4—HAE UAV Program Schedule As of 1998

- Re-structure schedule change (complete ACTD within funds available)
  - ACTD concludes December 1999
  - Reduction in AVs and ground stations
  - Reduction in Phase III period

mental flight-testing will not be complete by the time transferral occurs. As of July 1998, program plans included a continuation of such testing through January 1999, at which time the user demonstration-testing phase would commence. Thus, while Phase III has nominally been shortened by nine months (from 24 to 15 months' duration), in practice the user demonstration has been reduced 50 percent to 12 months.

Table 1.3 shows the earliest estimate of program costs for Tier II+ and Tier III–. The costs include both contract costs and government administrative and support costs.[19] Funding plans go through FY99, the end of the planned ACTD program. Future funding would be part of Phase IV production that would occur after the planned force-mix decision at the end of the ACTD. Program cost risk was considered to be contained by both the UFP constraint and a cap on NRE funding. The program risk was perceived not as exceeding cost but as trading performance.

While development costs grew significantly in both the Global Hawk and DarkStar programs, total program cost remained close to the original estimate. Table 1.4 shows a recent estimate of total budget and funding profile by program segment.[20] Funding profiles have

Table 1.3

HAE UAV Program Funding Plan
(in then-year $M)

Program	FY94	FY95	FY96	FY97	FY98	FY99	FY00	FY01	Total
Tier II+	23	82	140	150	150	150	TBD	TBD	695
									(676)
Tier III–	30	62	50	45	15	15	TBD	TBD	217
									(226)
Total	53	144	190	195	165	165	TBD	TBD	912
									(902)

SOURCE: JROC briefing, 8 Nov 94; *HAE UAV ACTD Management Plan, Version 1.0*, 15 December 94; and JPO data.

NOTE: Revised totals in parentheses are from a JPO briefing to the Association of Unmanned Vehicle Systems conference in Washington, D.C., 11 July 1995.

---

[19]Tier II+ budget line includes CGS Tier III– functionality, miscellaneous support, government-furnished equipment, studies, and the Phase-III demonstration costs for both Tier II+ and Tier III–. These figures are from "High Altitude Endurance Unmanned Aerial Vehicle Systems—Program Briefing for Joint Requirements Oversight Council," 8 November 1994, and are also found in a different format in the *HAE UAV ACTD Management Plan (draft), version 1.0*, dated 15 December 1994. Revised totals are from a JPO briefing to the Association for Unmanned Vehicle Systems conference in Washington, D.C., on 11 July 1995—FY breakdown is unavailable.

[20]The cost of including Tier III– functionality in the Tier II+ ground segment, contained in the Tier II+ budget line in Table 1.3, is contained in the Tier III– budget line in more recent budget documents, including Table 1.4.

Table 1.4

HAE UAV Program Funding
(in then-year $M)

Program	FY94	FY95	FY96	FY97	FY98	FY99	FY00	Total
Tier II+	10.7	68.3	102.3	70.3	99.8	82.2	10.5	444.1
Tier III–	42.4	57.1	85.9	64.8	43.9	33.6	6.0	333.7
CGS			1.2	51.2	47.8	39.8	9.5	149.5
Total	53.1	125.4	189.4	186.3	202.0	163.4	27.4	947.0

SOURCE: JPO budget charts, Jan. 1997. The *ACTD Management Plans* dated Aug. 1996 and Dec. 1997 show similar funding information.

changed from the initial plan, and the initiation of the CGS segment caused a reallocation of funds from Tier II+ to CGS, but the program is currently estimated to come in close to its original budget plan. This was accomplished, however, by dropping some NRE activities during Phase II, reducing test activities in Phase II and Phase III, and reducing the number of air vehicles and ground segments fabricated during the ACTD. The ACTD program now includes a maximum of five Global Hawks, four DarkStars, and two CGSs.

Though the program apparently will meet the budget, the scope of activities will be significantly reduced from the original plan. This means that the technical maturity of the systems—an understanding of the systems and their capabilities—will be less than originally anticipated, and the amount and type of information generated in support of the military utility and force-mix decisions will be reduced.[21] These are both significant nonmonetary costs that are not currently accounted for, and they raise the serious issue of achieving budget and schedules while not achieving the ACTD's objectives. According to one recent estimate, adequately completing the test and demonstration activities requires an additional $88

---

[21]While the Phase II activities for the Global Hawk will result in about the expected technical maturity, the Phase III test and demonstration activities have been significantly reduced. For Global Hawk, Phase II will not test radar Ground Moving Target Indicator (GMTI) performance or the survivability suite. This outcome—reduced technical maturity and information to support military-utility decisions—is somewhat worse for DarkStar.

million in FY00 funding and an extension of Phase III through September 2000.[22]

ORGANIZATION OF REPORT

This report covers the HAE UAV program from its inception through the end of Phase II. Because the operational test (Phase III) has not been completed, we are unable to analyze program outcomes definitively; we can, however, describe program formulation and engineering development, and we can characterize some of the ways in which contractors have responded to the special provisions of the program. In Chapter Two we provide an expanded discussion of the special acquisition approach being used in the program, as well as other issues to consider when analyzing an acquisition strategy. Chapters Three through Five summarize the evolution of the three components of the HAE UAV program: the Global Hawk, the DarkStar, and the Common Ground Segment (CGS). Chapter Six summarizes the perspectives of the Phase II contractors. In Chapter Seven we discuss the complexities of interprogram comparisons and compare the cost and schedules of related programs to provide a basis for evaluating results of the HAE UAV programs. In Chapter Eight we offer interim and provisional observations on how those programs have been affected by the special acquisition environment.

---

[22]See "High Altitude Endurance Unmanned Aerial Vehicle POM 00 Brief," 5 March 98.

Chapter Two

ANALYZING THE ACQUISITION STRATEGY

A COMMENT ON ACQUISITION STRATEGY

Because problems with UAVs were generally attributed to the acquisition process, rather than to inherent flaws in the UAV concept, the newer UAV programs became a logical target for acquisition reformers.[1] The specific set of reforms ultimately applied to the HAE UAV program, and the implementation of those reforms, evolved over roughly two years. The reforms were to provide a higher assurance of conducting a "successful" program; one that led to enhanced operational capabilities in a way that the users deemed to be worth the cost.

The HAE UAV program acquisition strategy reflects the applied reforms. The strategy defines management processes used in the program (including contracting mechanisms, status tracking, and reporting); provides incentives to the contractor (either rewarding positive performance or penalizing negative performance); identifies key decision criteria; defines the schedule as well as major technical and decision milestones; and defines the overall relationship between the government and the contractor.

---

[1]Although the CR UAV program was designated a Defense Acquisition Pilot Program (under Public Law 101-510, Section 809, 10 USC 2436) by 1992, the reorganization of the UAV JPO and the consolidation of CR and SR effectively rendered this decision moot (*Unmanned Aerial Vehicles (UAV) Master Plan 1992*, DoD, 15 April 1992, p. 24). When the Endurance program split into the Medium Altitude Endurance (MAE) UAV and the HAE UAV, both were designated ACTDs. The HAE UAV was established as a DARPA program with DARO sponsorship, and a JPO was established outside the UAV JPO.

The acquisition strategy used in the HAE UAV program was innovative across the entire spectrum of activity. Four specific attributes form the core of the program's innovative approach:

- designation as an ACTD, which largely defined the overall goals of the program and its schedule

- use of DARPA's Section 845 Other Transactions Authority (OTA), which essentially eliminated all traditional acquisition regulations, encouraged management tailoring and commercial practices, and transferred design responsibility and management authority to the contractor

- use of the Integrated Product and Process Development (IPPD) approach, which included use of contractor Integrated Product Teams (IPTs) with government representation

- defining a single requirement, UFP, and stating all other performance characteristics as goals which can be traded off to meet this target.

The HAE UAV program was the first program to attempt to combine all these innovations in an integrated acquisition strategy.

This research assesses the extent to which these acquisition-strategy innovations had their intended effect on both traditional and nontraditional program outcomes. Traditional outcomes include meeting cost, schedule, and performance goals; the familiar metrics are cost growth, schedule slip, and performance shortfalls.[2] However, while the innovative elements of the acquisition strategy used in the HAE UAV program are intended to affect traditional outcomes, they are also intended to affect other, nontraditional measures of program performance. These nontraditional measures of performance are increasingly important in the current acquisition environment.[3] They include: improved government-industry interactions; improved

---

[2]For examples, see Drezner, J. A., et al., *An Analysis of Weapon System Cost Growth*, Santa Monica, Calif.: RAND, MR-291-PAF, 1993; Rich, Michael, and Edmund Dews, *Improving the Military Acquisition Process: Lessons from Rand Research*, Santa Monica, Calif.: RAND, R-3373-AF/RC, February 1986.

[3]See, for instance, then–Secretary of Defense William Perry's vision in *Acquisition Reform: A Mandate for Change*, 9 February 1994.

communications; more flexible, responsive processes; faster de-ployment of new capabilities; and increased user involvement. Thus, in evaluating the acquisition strategy used in the HAE UAV program, both sets of outcomes must be considered.

It is also important to acknowledge that the acquisition strategy may not be the dominant influence on either set of program outcomes. Other factors can strongly influence outcomes, or the implementa-tion of acquisition strategy. These factors include the routine under-estimation of software development and integration tasks in recent weapon acquisition programs; underfunded engineering develop-ment, particularly early in the program; a sole-source environment; and the original source-selection decision that determines the set of industry organizations and capabilities that will interact with the government.

In other words, our analysis of the effectiveness of the HAE UAV ac-quisition strategy must distinguish between innovative and tradi-tional elements of the strategy; between acquisition strategy and other factors affecting outcomes; and between traditional and non-traditional outcomes.

INNOVATIVE ELEMENTS OF THE HAE UAV PROGRAM ACQUISITION STRATEGY[4]

ACTD Designation[5]

The ACTD process evolved in 1994 in response to recommendations of the Packard Commission (1986) and the DSB (1987, 1990, 1991).[6]

---

[4]This section draws heavily on RAND's Phase I report: Sommer et al., *The Global Hawk Unmanned Aerial Vehicle Acquisition Process: A Summary of Phase I Experience,* Santa Monica, Calif.: RAND, MR-809-DARPA, 1997.

[5]ACTD concepts and implementation are fairly complex, and the simple treatment here cannot adequately reflect this complexity. For more detail, see the ACTD infor-mation in the Acquisition Deskbook, and Tom Perdue, DUSD(AT), *Advanced Concept Technology Demonstrations (ACTDs),* and *Transition of Advanced Concept Technology Demonstrators (ACTDs) to the Formal Acquisition Process,* Final Draft, 29 August 1996. Also available on the OUSD(AT) web site (www.acq.osd.mil/at/).

[6]The material in this and the remaining paragraphs on ACTDs is drawn from the April 1995 *ACTD Master Plan* published by the DoD, and from a May 1995 summary by Hicks and Associates, Inc.

ACTD programs are intended to "quickly provide a warfighter with a prototype capability, to allow him to use that capability in realistic operational scenarios, to refine the operational requirements, to develop the concept of operations, and then to make a sound determination of the military utility prior to a decision to acquire."[7] ACTDs address identified military needs, often including formal memoranda from the JROC. The core elements of the ACTD initiative include an accelerated schedule (four to six years, including the two-year user-demonstration phase), early and continuous user involvement, integration of mature technologies, minimal residual capability left with the user at the end of the ACTD, and parallel development of both requirements and CONOPS. ACTD programs are more than demonstrations of technical feasibility, but are less than MDAPs in terms of production and support planning and activities.

ACTDs are defined as a "pre-acquisition" activity. Three basic options exist at the end of an ACTD program, depending on the outcome of the military-utility evaluation: terminate the program with no further work, continue development, or increase procurement of systems. If further development is required to meet the user's needs, the program might continue as an ACTD or enter the formal acquisition process at engineering and manufacturing development (EMD). If no further development is required and procurement of significant numbers of systems is warranted, the program might enter the formal acquisition process at an equivalent low-rate initial production (LRIP) phase.

ACTDs have a streamlined management structure, with the program director reporting to an Oversight Group chaired by the Deputy Undersecretary of Defense (Advanced Technology) and including all relevant decisionmakers from both acquisition and operational communities. Only two formal documents are required: a short Implementation Directive that broadly describes the ACTD and clarifies the roles and responsibilities of the agencies involved; and the Management Plan, a flexible and changing document that describes the acquisition strategy in detail, as well as sets up cost and schedule baselines. The Management Plan should include an ex-

---

[7] Perdue, "Advanced Concept Technology Demonstrations (ACTDs)," and "Transition of Advanced Concept Technology Demonstrators (ACTDs) to the Formal Acquisition Process," 1996.

plicit definition of military utility and how such utility will be demonstrated, although a description of a methodology for defining and demonstrating military utility later in the program may suffice.

The basic organization and acquisition strategy of the HAE UAV program evolved during the same period that the ACTD process was being formulated, and there was close coordination between the principals of both activities during 1993–1994. Therefore, while the HAE UAV program was formally initiated in April 1994 (solicitation to industry for Phase I submissions), before the ACTD process was formally introduced, it had a multiphase structure that was fully compatible with the ACTD process, and it was included with the projects that comprised the initial ACTD portfolio. A formal MoU designating the HAE UAV program as an ACTD was issued in October 1994.

One major consequence of designating the HAE UAV program as an ACTD was that the program could be started without going through the elaborate process required for typical Acquisition Category (ACAT) I and ACAT II programs, described in DODD 5000.1 and DoD Instruction 5000.2. Those traditional management procedures are based on the assumption that the new system will be produced and employed in significant numbers and well-understood ways, thus justifying extensive front-end planning and coordination. An ACTD program, however, offers an opportunity for radically new system concepts to be developed through a process whose tactics are developed along with the hardware, and the system's overall effectiveness is not judged until operational trials. Thus, program start requires less front-end planning and coordination, and critical decisions are deferred until demonstrated performance capabilities are available. ACTDs are explicitly not acquisition programs; an ACTD should quickly demonstrate military utility, not acquire many systems.

Section 845 OTA[8]

Another major element of the strategy to simplify the management process was use of provisions in recent legislation that permit re-

---

[8]The legislative history, intent, and concepts underlying the origins and use of Section 845 OTA are described in detail in Secretary of Defense Memorandum by Paul G.

moval of some oversight and management strictures typically found in government acquisitions. The HAE UAV program was designated a Pilot Acquisition program under the provisions of Public Law 101-189, 10 U.S.C. 2371, and Section 845 of the 1994 National Defense Authorizations Act (Public Law 103-160). This allowed DARPA to use an "agreement" in lieu of a contract, and permitted the waiver of Federal Acquisition Regulations (FARs), the Defense FAR Supplement (DFARS), the Armed Services Procurement Act, the Competition in Contracting Act, and the Truth in Negotiations Act (TINA), in addition to releasing the contractor from military-specification compliance. All procurement system regulations were inapplicable. It also freed the contractor from undergoing Defense Contract Auditing Agency (DCAA) audits, allowing instead the use of commercial auditors.

This authority is known as Section 845 OTA or, simply "DARPA Agreements Authority." DARPA already had the authority (under Section 2371) to use "cooperative agreements and other transactions" to implement its dual-use projects that feature cost-sharing with industry; DARPA had already implemented a few dozen of these "nonprocurement" agreements.[9] Section 845 expanded that authority to prototype projects that are directly relevant to weapons systems (i.e., those that are not dual use). HAE UAV was the first program to implement Section 845 authority.

Note that these Pilot Acquisition waivers were initially granted to HAE UAV only through Phase II (literally, for a period of three years from the enactment of the act granting the waivers). The extension of the waivers into Phase III was not assured at the time of Phase I and represented significant uncertainty for the bidding contractors. Section 804 of the FY97 National Defense Authorization Act (PL-104-

---

Kaminski, 14 December 1996, "10 U.S.C. 2371, Section 845, Authority to Carry Out Certain Prototype Projects"; Richard Dunn, DARPA General Counsel, Memorandum of Law, 24 October 1996, "Scope of Section 845 Prototype Authority"; Testimony of Richard L. Dunn before the Committees on Science, U.S. House of Representatives, 8 November 1995, "Innovations in Government Contracting Using the Authority to Enter Into 'Other Transactions' With Industry"; and Richard Dunn, "DARPA Turns to Other Transactions," *Aerospace America*, October 1996.

[9]For a review of experience with other projects, see Nash, M. S. et al., *Participant Views of Advanced Research Projects Agency "Other Transactions,"* Alexandria, Va.: Institute for Defense Analyses, Report D-1793, November 1995.

201) extended this authority through September 30, 1999, and also extended it to the military services and other defense agencies.

Section 845 OTA is intended to facilitate the use of commercial process and practices in weapons-system acquisition. It provides considerable flexibility in negotiating the terms and conditions of contractual relationships between the government and industry; it essentially allows the government to begin with a clean slate, with a minimum of preconditions. Specific benefits tend to fall into two categories of provisions: financial-management provisions that include access to records, accounting procedures, reports, and auditing; and intellectual-property provisions that include issues of technical data rights and patents.

Agreements under Section 845 OTA are fundamentally different from traditional contracts, though two FAR clauses are retained in all DARPA agreements: Article XI, "Officials not to Benefit," and Article XII, "Civil Rights Act," are required by law. Key differences from a typical fixed-price contract include Article IV, "Payable Event Schedule" (parties can agree that payable milestones can be altered based on program events), and Article VII, "Disputes" (the DARPA director is the ultimate arbiter of disputes).[10]

In practice, Section 845 OTA results in the transfer of substantial design responsibility and management authority to the contractor. Traditional oversight is significantly reduced and the government has fewer and less powerful mechanisms to directly affect the contractor's behavior, design, and other engineering activities. Other elements of the acquisition strategy—small JPO, IPPD/IPT process, and the streamlining associated with the ACTD designation—enforce this transfer. Given this transfer of responsibility, the contractors are encouraged to use their own management processes in the program.

It is recognized that program execution under Section 845 OTA can increase the risk to the government because of reduced oversight. Most of the government's experience with OTA is with dual-use projects involving multifirm consortia. In these projects, the risk is reduced by the contractors' interests in success (commercial applica-

---

[10]This material is largely drawn from the "HAE UAV Industry Briefing" given by the JPO on 5 May 1994.

tions), cost sharing, and self-policing by the consortia. For the most part, these conditions do not apply to the HAE UAV program.

Integrated Product and Process Development

An additional feature of the program is that the JPO strongly encouraged use of IPPD and associated IPTs. The heavy reliance on IPPD, while certainly not unique to this program, is a factor that must be considered when evaluating the effects of streamlined acquisition.[11] Use of IPTs is also strongly encouraged as part of ACTD program execution.

IPPD is defined in ACTD policy guidance as a management technique that simultaneously integrates all essential acquisition activities through the use of multidisciplinary teams to optimize design, manufacturing, and supportability processes. The concept has grown out of the practice of "concurrent engineering" and was first implemented in DoD by the Air Force during the F-22 program. On 10 May 1995, the Secretary of Defense directed that the concepts of IPPD and IPTs be applied to the acquisition process to the maximum extent practicable, in particular to the Office of the Secretary of Defense (OSD) oversight process.[12]

This mandate has led to two distinct implementations of IPPD. In OSD, the "vertical" IPT is a replacement for the former time-consuming serial program review process that is characterized by continuous meetings at ever-higher levels. In the IPT mode, all decisionmakers attend each meeting, and they expedite their decisions. IPPD implementation differs at the program level and more resembles industry practice. IPTs are formed in distinct product areas (that differ between programs) and are characterized by participants empowered and authorized to make commitments for the functional area or organization they represent. Key personnel are involved at an early stage, and timely decisionmaking is encouraged.

---

[11]The details of the UFP limit were finalized and the decision to implement IPPD was made during a series of trade-off studies and key meetings in the winter of 1994.

[12]Secretary of Defense William Perry, "Use of Integrated Product and Process Development and Integrated Product Teams in DoD Acquisition," letter dated 10 May 1995.

Government program-office personnel join contractor employees in IPTs, fostering teamwork and mutual trust based on shared data. Conceptually, the resulting free-flow of information should allow the government to know of problems earlier than in its traditional oversight role. The government defines performance objectives, rather than dictating product and processes, and imposes minimum oversight consistent with stewardship of public funds.

IPPD implementations vary widely across organizations. Since IPPD itself encourages tailoring IPTs to programs, IPPD has at times become a theme rather than a method, and thus difficult to analyze. The implementation of IPPD in the HAE UAV Tier II+ program took the form of a de facto mandate, although DARPA's stated intent was to *encourage* firms to organize as IPTs rather than by functional discipline. Nevertheless, the IPTs in the HAE UAV program did not have the full spectrum of functions continuously represented. That level of integration would have been expensive.

Unit Flyaway Price and Cost-Performance Trades

Improved cost control has been an important theme of acquisition initiatives for several decades. The results of initiatives such as "design-to-cost" and budgeting for risk have been mixed. The latest initiative is "Cost as an Independent Variable" (CAIV).[13] CAIV intends that cost be treated as an outcome variable in a way similar to the way performance has always been treated. The result is the occasional need to make difficult cost-performance trades in the design of a weapon system.

The HAE UAV Program Office introduced a radically different, and potentially stronger, method for controlling cost by treating it as the *only* required program "deliverable," *with all other performance objectives subject to trade-offs to meet the price objective.* Following program initiation, but prior to the award of Phase I agreements, the DUSD (AT) imposed a $10-million (FY94 dollars) UFP cap on both the Tier II+ and Tier III– programs. The $10-million UFP is defined

---

[13]USD (AT) Memorandum, SUBJECT: Policy on Cost-Performance Trade-Offs, 19 July 1995. Note that the "independent" nomenclature used in the CAIV policy is misleading. The intent of the policy is to treat cost as a dependent variable (outcome) or program objective, rather than as an input.

as the average price of air vehicles 11–20, including the sensor pay-load, in both the Tier II+ and Tier III– programs.[14] The UFP requirement differs from CAIV in that the contractor, not the government, is responsible for determining appropriate trades and the government has fixed the product price, not the trade space.

The degree of innovation that this approach represents can hardly be overstated. It is contrary to the established culture of acquisition management, where system performance has been the dominant criterion on which program success was based. Such a performance-dominated style has been well understood and practiced by both government and industry for several decades.

To accomplish this cost-control strategy—specifying only one firm requirement, the UFP discussed above—other desired performance characteristics were defined in terms of a range of values deemed acceptable, and the contractor was to find a balance among the various performance parameters so that the overall system satisfied the user's needs. This freed the JPO from monitoring the contractor's progress toward achieving individual performance specifications.

OTHER ELEMENTS OF THE ACQUISITION STRATEGY

Small JPO

In keeping with the plan to encourage the industry team members to organize and perform efficiently, the JPO was an austere organization. We cannot define the exact JPO staff levels during Phase I because they fluctuated; specialists were drawn from various agencies as needed. However, during most of Phase I the JPO apparently consisted of a core of about a dozen staff, plus another two dozen full-time-equivalent (FTE) specialists and support personnel. The size of the JPO during Phase II was approximately the same, about 30

---

[14]The $10-million UFP includes all flight hardware: airframe, avionics, sensors, communications, integration, and checkout. It is to be the total price paid by the government, including profit. Specifically, the UFP is defined as the average price for a Phase IV lot of 10 air vehicles, to be delivered over a 12-month period. Thus, the UFP limit is a projection, *not* a guarantee in the normal contractual sense.

FTE staff, including systems engineering and technical analysis (SETA) contractor support. Most of the JPO personnel worked Tier II+; Tier III– staff consisted of the deputy program manager for Tier III– and perhaps another half-dozen individuals brought in as needed on a part-time basis. The size of the Tier III– team increased after the April 1996 accident to approximately 15 FTEs, with an additional five to ten specialists brought in as needed.

Design for Low Risk

Another management strategy—one that was poorly defined during the early parts of the program—was for the JPO to design a program with relatively low risk of failure. This presented a challenging goal because traditional DARPA programs have emphasized high system-performance goals while accepting the concomitant risk of technical failure. Thus, the JPO's target represented a shift in emphasis. More specifically, the office seemed to believe that a program carrying low technical risk equated to one carrying a low risk of not achieving the UFP goal. To help achieve this goal, the JPO sent a strong message to the contractors that the development funds were limited to a specific amount, in hope that the contractors' plans and actions would be organized accordingly. In retrospect, the office believes the message was less clear than it hoped; at least one contractor during Phase I misinterpreted the JPO vision of how risks and other program objectives should be balanced. The JPO did not try to develop specific priorities among the major program goals (system performance versus UFP versus NRE versus risk of a major failure, etc.). Had it done so, its messages to industry during Phase I might have been more effective.

The goal of low overall program risk had a major consequence when the funding was cut prior to Phase II. In that circumstance, a traditional approach might have been to retain both contractors for competitive reasons and save money by reducing the amount of system maturity desired at the beginning of the Phase III operational test phase. However, the possibility of an unsuccessful test because of system-development deficiencies was too great, so the funding cut was accomplished by eliminating one of the contractors during Phase II.

SUMMARY OF DIFFERENCES

Table 2.1 summarizes the fundamental differences between a traditional acquisition process under the DoD 5000 series of regulations and the process used in the HAE UAV program, characterized by the ACTD designation and use of Section 845 OTA. Of special note is the

Table 2.1

Acquisition Process Comparison Framework

Factor	Traditional Process	ACTD/§845 Process
Purpose of process	Replace systems/force structure	Introduce new capabilities/concepts
Size of procurement (#, $)	Large	Small (niche)
Degree to which requirements are formalized (e.g., "ilities," performance)	Prior to program start; in detail; limited trade space	Loose, with more detail as knowledge is gained; larger solution space; larger trade space and flexibility
User participation	Low; requirements only	High; program structure, management, continuous evaluation and input
Process requirements (reporting, budgeting, milestones, etc.)	Standardized; formalized; oversight	Tailored; informal; "insight"
Maturity of technology	Immature; high risk	Mature
Nature of technical challenge	Full-spectrum development risk	Integration; CONOPS
Attributes associated with system type	Established constituency	No constituency
Agency operating environment	Mainstream SOP, culture	Joint; DARPA–other transaction
Institutional structure	PM —> PEO —> SAE/DAE PM —> functional staff	PM —> DUSD(A&T) —> DAE
Government-industry relationship	Capabilities drive funding	Funding drives capabilities
Testing	Dedicated test agency	Operational user w/support
Contractor organization	"DoD compliant"	Tailored, innovative business practices
Relationship to contractor	Adversarial	Cooperative
Maturity of mission concepts, CONOPS	Mature	Immature
Contractor design/management responsibility	Low	High

difference in purpose: the HAE UAV program was designed to quickly test new concepts and capabilities, and, if they prove useful, quickly introduce them to the operational forces. Traditional acquisition programs are generally intended to replace existing systems with a similar system having upgraded capabilities. Also, traditional programs generally involve large procurement quantities and the full spectrum of development technical risks. ACTD programs are intended to use mature technology, with the highest risk in the integration process.

Chapter Three

TIER II+ GLOBAL HAWK

The Global Hawk is the conventional configuration HAE UAV being developed by Teledyne Ryan Aerospace (TRA). Despite the fact it is intended to integrate existing and mature technology, it is a new design (not a modification) and its operational concept includes some challenging performance goals (fully autonomous control, ultra-long range and endurance). Additionally, the HAE UAV program was the first program to combine the previously discussed set of innovative acquisition-process attributes. From the perspective of the Tier II+ program, there was no precedent or prior experience on which industry and JPO decisionmakers could rely.

DESCRIPTION OF ORIGINAL PLAN

The basic concept for the Tier II+ program is a system capable of overt, continuous, all-weather, day/night, wide-area reconnaissance. The system is composed of three parts: an air vehicle segment, a ground segment, and a support segment. Only a few performance objectives were identified for the overall system. The flight vehicle was to be able to cruise to a target area 3000 miles distant, loiter over the target for 24 hours at an altitude of 65,000 ft, and then return to the take-off point. A mission-equipment package was to consist of an SAR and an EO/IR sensor, a data recorder subsystem, a threat-warning receiver subsystem, and an airborne data-link subsystem that would transmit data to the ground station that, in turn, would synthesize and display the sensor data.[1]

Contrary to typical practice, those performance characteristics were not mandated; all were listed as goals that could be traded against the one system characteristic that was a firm requirement. That single dominant requirement was that the flight segment had to be produced at a UFP not to exceed $10 million (FY94 dollars) for air vehicles 11–20.

Early in the program, it was DARPA's opinion that the complete set of *all* performance objectives could probably not be packaged into a $10-million UFP air vehicle. The contractor was to meet, or come close to meeting, as many as possible of the other system performance goals, but only the price limit was mandated. Ground and support segments, while not part of the UFP objective, had to be balanced in cost and capability.

Despite the emphasis placed on meeting the "required" UFP, the program was clearly structured so that a slightly different outcome could be accepted. In fact, it would have been unreasonable to state at the program start that no configuration costing more than $10 million would be accepted, especially given the meager information available at that time. The program organizers recognized that a preferred mix of price and performance might cost somewhat more than $10 million and still provide military value worth the investment. Thus, they inserted the instruction to the contractor to provide an affordable and reasonable future-growth path to meet all performance objectives, in case those objectives could not all be met with a UFP of $10 million or less. They expected competition to play a key role in constraining the final system price.

The Tier II+ program consists of four phases, as depicted in Figure 3.1.[2] Phase I was a six-month competition enacted between October 1994 and March 1995 that sufficiently defined the air vehicle, ground,

---

[1]While the solicitation provided the option of carrying either or both of the major sensor packages, the same section carried the statement: "As an objective, the air vehicle will have sufficient capacity to carry all prime mission equipment simultaneously."

[2]From Tier II+ Phase I Solicitation, June 1994.

RAND*MR1054-3.1*

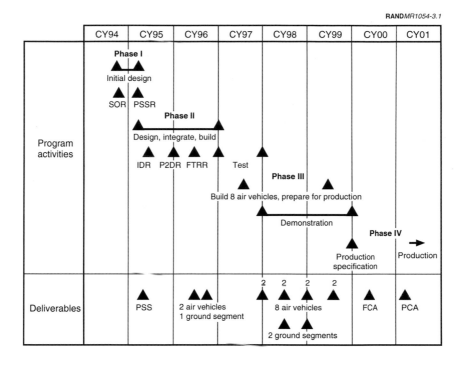

Figure 3.1—Original Tier II+ Program Schedule

and support systems, and system interfaces to provide confidence in achieving the performance goals within the UFP cap. That phase resulted in a preliminary system specification, a system-segment specification, and a proposed agreement to cover Phase II.

The Phase II plan was a competitive development consisting of two contractor teams designing and building two complete air vehicles (including payloads) and one ground segment each, together with flight tests sufficient to demonstrate technical performance and provide continuing confidence in the ability to meet the UFP limit. This phase was originally estimated to take 21 months to first flight, plus six months of development flight tests, but six months were added to the Phase II flight-test schedule by agreement with the con-

tractor; the Phase I Solicitation made it clear that the winning contractor would determine the exact length of subsequent phases.
Phase II was conducted using an updated agreement from Phase I
with a cost-type payment arrangement. A "thumbs-up" by the user
to continue and a system specification that all participants believed
would meet the $10-million UFP would signal a successful end to
Phase II.

Phase II was planned to end in December 1997. The HAE UAV ACTD
Management Plan identified four preliminary criteria for transition
from Phase II to Phase III:

• acceptable performance throughout the flight testing

• complete "end-to-end" sensor flight tests

• JPO at Wright-Patterson Air Force Base (WPAFB) manned

• HAE oversight group approval.

Phase III planned for the winning contractor team to build up to
eight additional air vehicles, two additional ground segments, and
support a two-year field demonstration of operational capabilities.
Management was to transfer to a joint service organization to be located in the Air Force Aeronautical Systems Center at WPAFB, and
headed by the Air Force, in mid-1997. The objective of Phase III was
a successful operational demonstration and completion of all tasks
that would achieve the $10-million UFP in Phase IV. Phase III's
planned duration was 30 months, though most program documentation shows a 24-month user demonstration, ending in December
1999. It would be conducted under an updated DARPA agreement,
and would mark the completion of the ACTD portion of the planned
program.

The plan envisioned the winning contractor making an "irrevocable"
offer of 10 air vehicles in Phase IV (beyond the 10 planned to be fabricated in earlier phases) for a UFP of $10 million (FY94 dollars) prior
to entering Phase III.[3] Providing Phases II and III were successfully

---

[3]See *HAE UAV Acquisition Management Plan*, December 1994. While adhering to the
UFP target, the Phase II Solicitation does not include the "irrevocable offer" language.

completed, Phase IV would consist of serial production necessary to meet operational needs.

The JPO issued the draft solicitation for Phase I on 28 April 1994. It described some aspects of the innovative management process that were enhanced at the May 1994 Industry Briefing, with emphasis on the UFP requirement and the implications of Section 845 OTA. The charts make it clear that the acquisition strategy's main feature was a significantly increased industry design and management authority and a more cooperative relationship with the government.[4]

SUMMARY OF PHASE I EXPERIENCE[5]

The solicitation for Phase I, issued June 1, 1994, stated the intent to select three firms, each of which would receive an "agreement" funded at $4 million. The funding was provided on a "not to exceed" basis, with payments based on completion of identified payable milestones.

The Phase I Solicitation made it clear that the $10-million UFP was the primary objective, and only requirement, for the program. The solicitation gave the contractors complete discretion to define the Tier II+ system, based on the performance goals described in the System Capability Document (SCD). The solicitation also specified that the contractors controlled the entire trade space.

Fourteen organizations, each consisting of a consortium of two or more firms, responded. This was an unexpectedly large response and included several nontraditional firms (such as Aurora and Grob). The acquisition process waivers granted under Section 845 OTA catalyzed the participation by nontraditional DoD suppliers. The number of bids inundated the small JPO. The bids presented a wide range of size and performance for a $10-million UFP, raising the question of the credibility of the cost estimates. Given the breadth

---

[4]See "Welcome to the High Altitude Endurance Unmanned Aerial Vehicle (Tier II+) Industry Briefing," 5 May 1994.

[5]This material is based on information in RAND's previous report on Phase I experience, Sommer et al., MR-809-DARPA, 1997.

and quality of the responses, DARPA selected five teams to perform Phase I:[6]

- Loral Systems Co. with Frontier Systems, Inc.

- Northrop Grumman Aerospace Corp. with Westinghouse Electric Corp.

- Orbital Sciences Corp. with Westinghouse Electric Corp.

- Raytheon Co. Missile Systems Division with Lockheed Advanced Development Co.

- Teledyne Ryan Aeronautical with E-Systems Corp.

During Phase I, DARPA revised its plans for Phase II, which called for building two flight vehicles, one ground and control segment, and a system demonstration. Although DARPA had previously announced that two Phase II contractors were to be awarded agreements (and thus two systems designed and flown), funding limitations forced a down-selection to only one Phase II contractor. This early elimination of competition within the Tier II+ program proved to be controversial, both within contractor circles and on Capitol Hill. The initial Tier II+ funding plan and the revised plan are presented together in Table 3.1;[7] note that these totals represent funds obligated to the contractors, not total funds available to the program. DARPA did not alter the program plan further, despite the fact that a competitive Phase II was a major part of the overall program design. Therefore, Phase II would essentially be a single-source, cost-plus-incentive-fee (CPIF) activity, with the contractor obligated to deliver a best-effort design that could be produced for a unit price of $10 million. While DARPA would write incentive clauses into the agreement, the incentive of winning a subsequent production award on the basis of a flyoff against a competing firm had been eliminated.

----

[6]Only the principal airframe and electronic system members of the teams are listed here; most of the consortia included additional members that provided specialized services.

[7]Original figures are from the draft Tier II+ Solicitation dated 29 April 1994. The release version of the solicitation (ARPA PS 94-33) dated 1 June 1994 had a funding profile that was smaller by $10 million: $70 million instead of $75 million in both FY95 and Phase III of FY97 (not reflected in this figure). The revised numbers shown are from the Tier II+ Phase II solicitation, dated 15 February 1995.

Table 3.1

Tier II+ Program Obligation Plan
(in then-year $M)

	FY94	FY95	FY96	FY97	FY98	FY99	FY00	FY01	Total
Phase I									
(3 contractors)	(12)								(12)
5 contractors	20								20
Phase II									
(2 contractors)		(75)	(110)	(50)					(235)
1 contractor		40	80	38	6				164
Phase III									
1 contractor				(75)	(130)	(50)	(20)		(275)
1 contractor				55	94	99	0		248
Phase IV									
1 contractor						TBD	TBD	TBD	TBD

SOURCE: Original funding profile in parentheses from Tier II+ Phase I Solicitation.

The Phase II Solicitation (15 February 1995) stated two program objectives: (1) Produce an HAE UAV reconnaissance system that provides the most military utility for the $10-million UFP, and (2) execute the program as a model for future acquisitions. The solicitation also stated that cost-performance tradeoffs were critical to achieving both objectives. While DARPA changed the funding profile to reflect available funds (see Table 3.1), it made no changes to the nominal schedule (see Figure 3.1). A more detailed task description document (TDD), integrated master schedule (IMS), and integrated master plan (IMP) were required to provide visibility into all subsystems and define UFP allocations. The Phase II proposal was also to include a two-part Phase III option: a fixed price for an additional eight air vehicles, a cost-plus estimate for two more ground segments, and logistics support of tests. The solicitation again suggested a management approach that emphasized IPT structures, maximum use of commercial systems, streamlined processes, and contractor responsibility.

The TDD, IMP, and IMS were to use the same structure to enhance the transparency of their relationship. The Phase II Solicitation provided detailed insight into the preferred structure and function of the IMP. The IMP was to have two parts: Product and Process. The Product IMP is the version normally associated with program management, and defines the events that must occur to successfully design, develop, and test the system. The top level of the Product IMP would include the following minimum events: initial design review (IDR), final design review (FDR), flight readiness review (FRR), first flight (FF), Phase III production readiness review (PRR-3), Phase III demonstration readiness review (DRR), and Phase IV production readiness review (PRR-4). Each task was to include criteria to judge its completion. DARPA intended the Process IMP to allow the government to see into management processes and how those processes related to the products; examples include UFP tracking and software development processes. The Phase II IMS was intended to be traceable to the original IMS submitted in the Phase I proposal. Within these guidelines, the contractor was free to tailor its proposal.

As set out in the February 1995 Phase II solicitation, DARPA's evaluation of the five contractor proposals was to evaluate the proposed system capability, technical approach, management approach, and financial approach. The specific criteria were tied to the $10-million UFP, reemphasizing its importance.

PHASE II BASELINE

The TRA team won the Phase II award in May 1995. DARPA intended the selection of TRA to be relatively low risk. TRA delivered a relatively conservative air-vehicle design, and at the time of source selection, the technical risk was expected to derive from the flight segment. In contrast, TRA's design had a relatively high cost risk; its design was at the high end of the weight scale, implying more cost but lower technical and performance risk. An illustration of the successful TRA Global Hawk air-vehicle design is presented in Figure 3.2.[8]

The Phase II Agreement executed on 3 August 1995, with an effective date of 6 April 1995, was an amended version of the original agreement between DARPA and TRA covering Phase I. Reflecting the streamlined management process developed by DARPA for programs falling under the Section 845 OTA, the amended agreement was only ten pages long. During the first three months of the program, the

---

[8]TRA artist's concept.

Figure 3.2—Global Hawk Configuration

details of project design were modified and extended beyond that of the original TRA proposal. The agreement included by reference a System Specification, a TDD, an IMP, and an IMS. Together, those documents described the system capability goals and how the Phase II activities were to be organized.

Phase II activities included completing the Tier II+ design, defining the system specification and interfaces, fabricating two air vehicles and one ground segment, and completing an initial flight test program. To accomplish this, TRA included in the Agreement a list of guidelines and processes intended to ensure low risk development and high military utility:

- early testing

- compatibility with existing military systems

- integrated product development philosophies
- tradeoffs to maximize military utility
- built-in growth path
- maximized use of off-the-shelf (OTS) equipment
- maximized use of open architectures
- minimized system life cycle cost
- required supplier participation in the IPT structure
- invited customer participation in the IPT structure.

These guidelines reflect the management philosophy of the program, including user participation, IPT structure, performance tradeoffs, and use of OTS equipment.

The agreement for Phase II included a statement that "TRA has established strategic alliances with proven leaders from industry who possess the experience and expertise to make a significant contribution to the success of the Tier II Plus Program." It listed several firms:

- E-Systems, for the ground segment and the airborne self-protection suite
- Rockwell, for the composite wings
- Allison Engine, for the propulsion system
- Hughes Aircraft, for the radar and EO-IR sensor suite.

However, none of these firms was a signatory to the agreement. All arrangements between TRA and the supporting contractors were private and not part of the agreement between TRA and the government.

The TDD, dated 31 July 95, was included as Attachment 1 to the Phase II Agreement and included the IMP. Neither the Agreement nor the attachment explicitly distinguished the Product and Process IMPs. The TDD described the basic tasks for designing, developing, and testing the Global Hawk system. All tasks were related to the minimum-event list above, as well as additional events inserted by TRA (design review, element integration complete (EIC), flight-test readiness review, and flight test for the first payload flight). The TDD did not provide event dates, nor did it indicate the priority tasks along a critical path, except that presumed by the hierarchical structure of the TDD itself.

The negotiated Target Cost, Target Fee, and total CPIF Amount were:

Target Cost: $148,177,000

Target Fee: $9,171,000

CPIF Amount: $157,348,000

An additional cost-plus-fixed-fee (CPFF) amount of $640,315 was also included in the Phase II Agreement for other tasks. The target fee was 6.19 percent of target cost, with a maximum of 15 percent and a minimum of 1.19 percent. The actual fee would be determined within this range through an Earned Technical Performance Incentive defined in the Agreement. A 50/50 cost share of overruns was reflected in a fee reduction. TRA was obligated to perform work only until the value of that work equaled the funds committed by the government.

The agreement included, by reference, an IMS. While the details of that schedule, and the associated IMP, were modified through negotiation during the early months of Phase II, the key program milestones remained:

- Phase II start: April 1995

- First Flight: December 1996

- Phase II end: December 1997

The schedule was tight. A period of 20 months from development start to first flight is exceptionally short for a system as complex as the Global Hawk. While first flight did not require full integration of the payload, it required that the entire flight vehicle, the Launch and Recovery Element (LRE) of the ground-support system, the associated communications links, and all related software be fully developed and validated to a high confidence level, because the system was to operate autonomously from the beginning.

The Global Hawk system comprised the following main elements and sub-elements:

- Air Vehicle Segment, consisting of the flight vehicle and payload
- Ground Segment, consisting of the LRE and Mission Control Element (MCE)
- Support Segment, to provide the logistics necessary to use the system in the field.

As prime contractor, TRA was responsible for each of the three major segments and for their integration into the overall system. Program management was partitioned into six elements, each managed by an IPT:

- Air Vehicle Segment
- Payload Segment
- Ground Segment
- Support Segment
- Systems Engineering/Program Management
- System Test.

The DARPA JPO had a mirror-image organization so that, in theory, each segment leader in the JPO could work directly with his counterpart in the contractor's team.

For most of these system segments, the task definition and general development strategy remained unchanged throughout the program. The exception was the Ground Segment. By the time Phase II started, some envisioned a CGS to support both the DarkStar and Global Hawk systems. For the initial development of each system, each contractor (Lockheed for DarkStar; TRA for Global Hawk) was to design a specialized ground segment, anticipating that the two designs eventually would be merged into a common configuration. In practice, this meant including Tier III– functionality in the Tier II+ ground segment.

The close contractor-to-contractor relationship and information sharing required to make CGS work did not happen. To resolve the issues, DARPA took over management of the CGS in mid-1996 and contracted directly to E-Systems. That new CGS is scheduled to be available by mid-1999 to support both systems during Phase III operational demonstrations. Whether that CGS will accomplish common processing of image data, or whether the differences between the DarkStar and Global Hawk sensor configurations will result in separate image processors being incorporated into an otherwise common ground segment, is uncertain. We discuss this matter more fully in Chapter Five.

The decision on the CGS left each of the flight-system developers free to develop ground segments necessary to support their own system development and demonstration programs. Managing the development of the CGS was not formally an obligation of TRA and its funding was not included in the TRA Agreement covering Phase II. The eventual change in management, with Raytheon E-Systems taking the lead for CGS, required no change in the Phase II Agreement with TRA. However, it put the government in the role of system integrator for the three HAE UAV segments.

Other Attributes of the Agreement

The agreement under Section 845 OTA defines the government-industry relationship and has several characteristics that are substantially different from those of a traditional contract. These include: few contractual obligations (no detailed system specifications and deliverables); the contractor's ability to unilaterally stop work at any time without penalty; no provision for award protest; and limited government direction. Payable-event milestones defined by the contractor are incorporated. Additionally, no formal reporting or tracking processes or systems were mandated; TRA could use its own processes and work breakdown structure (WBS), which is reflected in the TDD, IMS, and IMP.

Limited government influence was specifically incorporated into the agreement:

> "This agreement gives extraordinary responsibility and authority to TRA. The Government will not unilaterally direct performance within or outside the scope of the work. Thus, the government must be able to convince TRA of the need for change."

This paragraph is unique to the TRA Phase II Agreement (Article XXVI); it means that TRA does not need to take direction from the government. This article emphasizes that TRA has primary responsibility for design and programmatic decisions, and full authority to make them. Given the nature of the IPT structure, and the intended relationship between government and industry, this limitation makes sense, though it increases the workload of government participants. It forces the government to build a consensus with the contractor, influencing TRA's decisions indirectly through IPT participation and information sharing.

The clause turned out to be a double-edged sword. While it inhibited the government from the tinkering and meddling common in other programs, it also prevented the government from imposing changes in contractor management structure and processes that were needed.

PHASE II EVENTS, OUTCOMES, AND EXPLANATIONS

Compared with Phase I, the Tier II+ program became increasingly complex as activities transitioned from conceptual and analytic studies to engineering development and associated hardware and software products. Below is a list of some of the main events and conditions that affected the Global Hawk program during Phase II.

- Reduction in budget leading to a loss of the competitive environment in Phase II. While the budget cut occurred in Phase I, it affected Phase II execution by radically changing the contractual and management environment from one that relied on competition to ensure contractor performance to a single-source best-effort arrangement with weak incentives and limited mechanisms for government intervention.

- Underestimation of integration risk. Inadequate emphasis was placed on the risks of software development and systems integration.

- The lead contractor (TRA) was a relatively small organization with good experience in small UAV programs but little experience in large, complex programs. TRA's primary expertise was in the air-vehicle system; it had inadequate capabilities in key software and integration areas. Being small, it had limited resources to apply to problems. It also had relatively weak management processes that, at least initially, were driven by personalities.

While these items are important, they do not tell the whole story. Nor, by themselves, do they enable an assessment of the viability of the acquisition strategy.

The major technical challenges were software development and systems integration. TRA did not acknowledge the need for significant new software development early in the phase; during the first year of Phase II, TRA did not have adequate software development and integration capability on the program. The government knew that TRA was relatively weak in this area at source selection, but because TRA was judged to be strong in air-vehicle design and development, the government was comfortable with its award to TRA. As work progressed, it became clear that the critical technical challenge was in system integration, not in air-vehicle design as originally anticipated. TRA's management refused to recognize the problem when the government pointed out this gap between program risks and contractor capabilities, but the government had little recourse due to the structure of the agreement. Over a year after the start of Phase II, a change in contractor management facilitated the application of appropriate resources, and the software development and integration problems were adequately addressed.

The acquisition strategy emphasized the use of COTS equipment as a way to give the contractor more freedom in design and control costs. However, the risks associated with integrating COTS into a complex system like the Global Hawk were underestimated. In some areas, such as the mission computer, the COTS equipment needed substantial redesign and development.

During the initial detailed design phase early in Phase II, the JPO communicated through the Air Vehicle IPT that it desired a significant redesign to strengthen the wing. TRA's analysis showed that the strengthening was unnecessary, but eventually made the redesign, incurring additional cost and time. TRA informally requested compensation for the change, viewing it as a government directive. The JPO felt that air-vehicle design was the contractor's responsibility and because TRA agreed to the change, no compensation was necessary. This disagreement was due in part to the lack of a formal engineering-change process, and in part to a clash of cultures and personalities as the operational details of IPT functioning were worked out. Dispute remains as to whether the strengthening was needed.[9]

Beyond the inadequate resources applied by the contractor to the software-development and system-integration challenges, an additional explanation for technical difficulties experienced in the Global Hawk program is that the technologies being used for both the air vehicle and payload systems were not as mature as expected. To the extent that the technology was immature, the program may have violated the ACTD criteria for the use of mature technologies, a policy intended to lower program risk and focus attention on integration.

Another core aspect of the ACTD process is the early participation of the user community. Military users apparently participated minimally during Phase II; they were informed of program status and capabilities, but provided little input to the development process. Thus, the opportunity the ACTD provided for users to affect development was passed by. This is not surprising, given that user input to engineering development is an unfamiliar role to both users and developers. The users are the focus of the Phase III demonstrations.

The April 1996 crash of the first DarkStar during takeoff on its second flight affected the Global Hawk (and CGS) program. Although the two projects are independent of each other, the DarkStar crash resulted in heightened risk-aversion throughout the HAE UAV program. This was expressed as more-conservative design decisions and increased reviews, as well as a previously unplanned single-point failure analysis and increased testing on the System Integration Lab (SIL) prior to first flight. The JPO and TRA added activities to Phase II to ensure adequate development and maturity prior to first flight.

Weak processes for managing the technical, cost, and schedule elements apparently affected outcomes significantly in the Global Hawk

---

[9]In an additional, unrelated technical event, the V-tail failed during static test. This was a surprise, but its effect on the program was not critical.

program.   Section 845 OTA allowed all traditional DoD systems-engineering processes to be waived;[10] TRA apparently did not substitute equivalent processes until late in Phase II. The result was an apparent lack of engineering discipline.

Early in the phase, the IMP and IMS were not kept integrated or up to date.   The engineers who were the technical leaders of the IPTs worked to schedule and cost targets, but did not adequately record their progress, i.e., track current cost and schedule.   TRA's new management structure adopted a method that overlay business concerns to ensure adequate cost and schedule tracking.   In April 1997, coinciding with the start of a new program manager, TRA implemented a process that fully integrates cost and schedule status into their earned-value system.   The IMP and IMS have been updated to reflect the current status of the program.[11]

Because of the paucity of program documentation, we cannot exactly determine when certain problems arose or attracted special management attention.   For example, the Phase II proposal included a Master Program Plan and an associated Master Program Schedule, albeit at a relatively rudimentary level of detail.   During the first three months of the program, interaction with the JPO modified the Master Program Plan, but the Master Program Schedule was never treated as a major element of program management.   It was not until mid-1996, about 15 months into the program, that JPO insistence and assistance led to implementation of a documented and maintained schedule system for tracking and managing various development tasks.   However, TRA did not regularly update the IMP and IMS before April 1997.   Thus, we cannot identify schedule deviations at a detailed task level until well into the second year of the program, and even then we cannot determine if a particular schedule deviation was the result of something that happened within that task or

_____

[10]For a summary of DoD's recommended systems-engineering processes, see Defense Systems Management College, *Systems Engineering Management Guide*, December 1986.

[11]Technical management of the program was facilitated by a Quality Assurance Report (QAR) database capable of incorporating all program related information. TRA generated QARs as problems or discrepancies were uncovered, allowing them to be tracked and resolved. The JPO had access to the database and could review the status of all QARs.

whether it was a consequence of some linked task experiencing delays. Given the difficulty of tracking the program even after the fact, neither TRA or JPO management could hardly have known the relative status of each program element. We can observe, however, that the nature of the relationship between TRA and the JPO within the IPT structure ensured that the JPO and the contractor had exactly the same information on program status at essentially the same time.

An additional observation regarding this schedule-tracking system is that it apparently was inconsistent over time.[12] At the detailed level, many activities listed in the initial versions are not listed in more-recent versions, and no indication exists of why they were dropped or how else they were incorporated. Similarly, these recent versions include items not originally listed, with no explanation of why they now merit more-formal tracking. We likewise see no way to determine which events are dependent on previous events, nor to determine the critical path. Name changes were common. One explanation for the changes is that the systems-engineering discipline normally enforced by mandated use of formal DoD procedures was missing and not replaced by equivalent TRA processes. The changes also may indicate a chaotic engineering-development program.

One other factor that surely affected the Phase II schedule was an apparent subtle shift in strategic emphasis that occurred during 1996. At the beginning of the program, the solicitation and the subsequent agreements for both Phase I and Phase II clearly specified that the paramount objective was a system design that could be produced for $10 million per copy. However, that objective was obviously contingent on successfully developing a system that could come close to meeting the mission-capability goals. Without a system whose performance could demonstrate reasonable accommodation of program goals, meeting the price goal would be irrelevant. As the program evolved and problems arose, management emphasis apparently shifted toward creating a system that could be successfully demonstrated. This is based on the observation that what little program documentation existed throughout 1996 and 1997 included little evidence of tracking design progress against the UFP goal, but included considerable evidence of tracking the engi-

---

[12]RAND has copies of the tracking-system output from July 1996 through April 1997.

neering progress toward achieving a design that could perform reliably (i.e., without crashing) and effectively. That goal became so paramount that the JPO insisted on full demonstration of the system design on the SIL, regardless of schedule slip, before flight testing could begin.

Key outcomes of the Global Hawk program to date are:

- Performance goals will likely be met.

- Government-industry interactions evolved into a positive, open working relationship, providing the JPO with timely insight into status and problems.

- Decisions are made faster because of increased contractor responsibility.

- Overhead costs are lower because of the elimination of complex reporting, oversight, and auditing.

- Reduced assets support a shortened user evaluation.

- The program experienced schedule slip, developmental cost growth, and an increase in the UFP.

Despite the technical problems experienced so far, program performance goals are likely to be met. The latest table of performance parameters provided by the JPO indicates that the current expected performance of the Global Hawk is essentially the same as indicated in Table 1.1.[13] No functionality has been dropped from the system, despite cost constraints, and the technical problems appear to have been satisfactorily resolved. As discussed below, by the end of May 1998, three substantially successful test flights had been conducted. Information generated from the flight-test program to date does not indicate a need to significantly revise expected performance.

However, two results of the technical problems were cost growth and schedule slip. First flight was delayed by 14 months from the original plan, and the user-evaluation phase was compressed in order to maintain the timing of the production decision. NRE costs have in-

_____
[13]See the full table on the HAE UAV program web site: www.darpa.mil/haeuav/charac.htm/.

creased. In addition, the quantity of vehicles procured prior to the production decision was reduced from ten to five. Note that both the schedule compression and quantity reduction for Phase III could be driven by cost and budget issues, as well as the technical problems; in a program review, the USD (A&T) directed that the system remain within available funding.

Figure 3.3 shows the evolution of major milestones for the Global Hawk based on available official program documentation through July 1998. Notable changes are the slip in first flight, the delay in program-management transferral to the Air Force as a result of the Phase II extension, and the compression of the user evaluation in Phase III. The December 1999 end date for the ACTD has not changed, despite changes in other key events. DARPA planned to transfer management to the Air Force in September 1998, marking the official end of Phase II, but development testing will continue into the first months of the Phase III user-demonstration period. The total amount of testing and user evaluation prior to the force-mix and production decisions at the end of Phase III has been reduced. Perhaps more importantly, the user evaluation has been effectively shortened to 12 months (January 1999 through December 1999) and will include a reduced set of operational scenarios and environments. One result is that the quality and quantity of information generated in support of the military utility and force-mix decisions has been greatly reduced. All program funding for the ACTD ends December 1999.

Drawing on schedule data from the system implemented in mid-1996, and available program documentation, one can track the evolution of a few major milestones. The plot of first-flight date, as the planned schedule evolved from month to month, is shown in Figure 3.4. The original plan scheduled the first flight to occur in December 1996. No specific rescheduling occurred until mid-1996, when the first elements of the new schedule-tracking system were introduced. At that time, first flight was anticipated in February 1997, a two-month slip. Thereafter, the first-flight date receded in near-real time for about a year.

Two factors appear primarily responsible for that continuing delay. First, the companion DarkStar program experienced a crash on its second test flight on 22 April 1996, and the investigation indicated

RAND *MR1054-3.3*

Global Hawk Schedule Evolution	Industry Brief	Phase II IMS	IMS/ IMP review	Design Review	Mngt Plan, Ver 6.1	Heber SASC testimony	Mngt Plan, Ver 7.0	DUSD (AT) web site	HAE UAV User Conf.	POM 00 Brief
Milestone	May 94	Mar 95	Jun 95	Jan 96	Aug 96	Apr 97	Dec 97	Feb 98	Feb 98	Mar 98
Phase I contract award/start					Oct 94		Oct 94			
Phase II contract award/start	Apr 95		May 95		Jun 95		Jun 95			
Initial design review	Aug 95	Aug 95	Aug 95	Aug 95	Aug 95		Aug 95			
Final design review–1		Nov 95	Jan 96	Jan 96	Jan 96		Jan 96			
Final design review–2	Dec 95	Mar 96	Apr 96	Apr 96	Jun 96		Oct 96			
EIC		Sep 96	Sep 96	Sep 96						
Flight test readiness review–1	Jun 96	Nov 96	Nov 96		Nov 96		Aug 97			
First flight	Dec 96	Dec 96	Dec 96	Dec 96	Nov 96	Nov 97	Nov 97			Feb 98
Flight test readiness review–2		Apr 97	Apr 97							
First flight w/payload		May 97	May 97	May 97	May 97		Feb 98			Aug 98
JPO transition to AF					Nov 97			Oct 98	Mar 98	
Phase II end/ Phase III contract award/start	Aug 97		Dec 97		Aug 97	Sep 98	Nov 98	Oct 98		
Phase III production readiness review		Jun 97	Jun 97							
Phase III demonstration readiness review		Nov 97	Oct 97							
Phase IV production readiness review		Nov 99								
Phase III initial military utility assessment					Aug 98		Aug 98			Jul 99
Phase III final military utility assessment					Nov 99		Feb 00			Jan 00
Phase III end	Dec 99	Dec 99			Feb 00	Sep 00	Feb 00		Dec 99	
Production/force mix decision					Feb 00		Feb 00	Oct 99	Dec 99	

NOTES: All dates are estimates unless date of estimate is earlier than value in cell. Source documents often provide estimates as quarters, e.g., 1QFY97. Midpoint in quarter is selected as rough estimate. Some dates include entire HAE UAV program (DarkStar and CGS): program office established; military utility assessments; production/force mix decision; JPO transition.

Figure 3.3—Evolution of Key Milestones for Global Hawk

that problems in flight-control software were the major cause. JPO managers became increasingly insistent that flight-control software be fully demonstrated on the SIL before the first flight. Second, problems in developing the software delayed that full demonstration

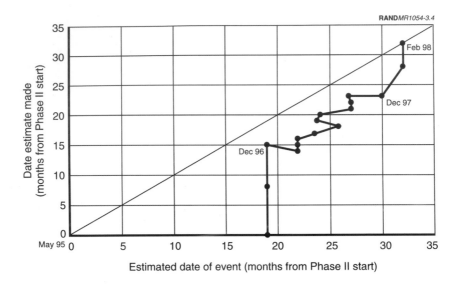

Figure 3.4—Evolution of First-Flight Estimates

on the SIL. The result was a continuous rescheduling of the first flight as the problems persisted. Delays in first flight were also caused by interface-definition problems, including command and control and data processing/links.[14]

Air Vehicle 1 rolled out on 20 February 1997. AV-1 was transferred to Edwards Air Force Base on 28 August 1998 and began taxi tests in October. The Flight Readiness Review occurred February 9, after which the JPO Director suggested that TRA take an extra week to perform additional confidence testing. Poor weather caused additional delay during February.

First flight of Global Hawk occurred on 28 February 1998 at Edwards AFB. While several problems arose during the flight, the system proved robust enough to handle the problems using manual-override commands that were not scheduled for testing until future flights.

An additional aspect of the first flight was that attendance was strictly controlled by the JPO director. Only program personnel observed

_____

[14]The pattern of slip in first-flight date is not correlated with external factors such as budget cycles or changes in contractor management.

the flight; the only VIP in attendance was the DARO Director. This strategy took pressure off the early flights; test flights are more likely to be canceled if there are no VIPs to disappoint.

A successful second flight of Global Hawk occurred on 10 May 1998, five weeks later than the planned date of 4 April 1998. Global Hawk's third flight occurred on May 30.

Figure 3.5 shows the evolution of Phase II costs for the Global Hawk. The total cost of the agreement includes all activities funded during Phase II, as well as the government portion of any cost share, contractor fee, and contractor cost share.[15] Total growth was 121 percent, approximately two-thirds of which is attributable to an increase in Phase II activities. The major cost increase for additional activities was in Phase IIB, which authorized fabrication of the third and fourth air vehicles and long-lead procurement for the fifth. Other activities included contractor-acquired property, support for several

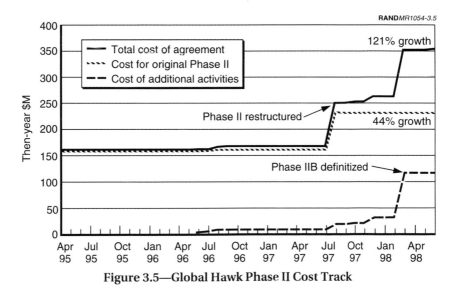

Figure 3.5—Global Hawk Phase II Cost Track

[15]The renegotiated Phase II Agreement (Amendment 24, 4 August 1997) states that the contractor fee may be used to offset contractor cost share for expenditures above the target price. Thus, it is not possible to track cost share precisely using information available in the agreement and amendments.

studies, and planning Phase III. With these activities removed, the cost growth (to the government) in the original statement of work is 44 percent, from $158 million to $226 million.[16]   Table A.1 in the Appendix presents this data in more detail.

The UFP is also likely to increase above the $10-million target, by an estimated $2–$3 million, for several reasons. The development cost grew because of technical problems, further software development, and difficulty integrating COTS. However, these are essentially NRE costs and could be allocated so that the UFP of vehicles 11–20 is not affected. More importantly, no functionality has been dropped from the system. The contractor has made few cost-performance trades, even with complete discretion to do so. The unwillingness to make such trades appears to result from the ambiguous definition of military utility. Because future production, beyond the nominal Phase IV plan, is important to TRA, the company was unwilling to drop functions that the users might want and rate highly during the evaluation. Additionally, there is a perceived competition for funds between DarkStar and Global Hawk that had the effect of impelling TRA to keep its system goals. TRA also claims that the JPO indicated that performance trades that degraded functionality would not be viewed positively, though the JPO denies sending that message.

Interestingly, the potential for a UFP breach was briefed up through the JPO to the highest levels in DARPA. While DARPA made no adjustment to the UFP, the process was unusually open, with a candid discussion of UFP issues between the government and the contractor.

As recently as August 1998, the contractor was discussing reducing payload functionality planned for the Phase II tests, and restoring it in Phase III. The discussions mainly involved smaller subsystems, such as the survivability suite; TRA maintained that integration of these subsystems during Phase II would take additional time and money. Delaying these nonrecurring activities would reduce the

---

[16]While many of the added activities are clearly outside the scope of the original statement of work, others are not so clear, including the cost of a government-directed one-week delay in first flight and some NRE tasks under the Phase IIB amendment. Thus, the 44-percent cost growth may understate the real growth needed to complete the original task.

amount of cost-share that the contractors have to pay by reducing the scope of Phase II activities. The program office observed that the contractors are "creating" additional NRE funding by trading functionality in Phase II, reducing developmental activity and thus lowering total costs and cost growth associated with Phase II. This is contrary to the intent of the acquisition strategy, which was that functionality be traded in order to maintain UFP. However, the structure of the Agreement allows both types of trades and does not provide a mechanism for the government to endorse one type of trade over another.

In some of the latest program documentation (POM 00 Brief, 5 March 1998), the JPO Director indicated that while the ACTD program formally ends December 1999 (funding stops), the program will not be complete. He recommended adding $88 million and nine months of additional testing to Phase III. This would significantly increase the Phase III user evaluation, improving both the quantity and quality of information generated through the demonstration program.[17]

Renegotiated Phase II Agreement[18]

The Global Hawk Phase II Agreement was renegotiated (signed 4 August 1997) to accommodate the problems experienced. Some of the technical performance measures (TPMs) were changed. The original agreement followed a CPIF contract approach. The new agreement requires cost-sharing at a threshold of $206 million of program cost at a ratio of 30 percent TRA, 70 percent government, until a value of $228 million is reached, where the program is capped. Previously earned fees must begin to be paid back to the government at that point. TRA's subcontractors begin participating in the cost share at $218 million. TRA is not obligated to continue to perform when the limit is reached unless the Agreement is further modified. The renegotiation also required that TRA and its team members invest $3.1 million in the SIL, above the value of the Agreement.

---

[17]See "High Altitude Endurance Unmanned Aerial Vehicle POM 00 Brief," 5 March 98.

[18]DARPA Agreement No. MDA972-95-3-0013, Amendment No. 0024, 4 August 97.

Phase IIB Amendment[19]

TRA and DARPA amended the Agreement, effective 31 March 98, to include activities associated with a "Phase IIB": provide three additional Global Hawk air vehicles, an integrated sensor suite, technical manuals, spare parts, training, and software maintenance. This amendment is intended to facilitate the transition into, and conduct of, Phase III.

The purpose of the Amendment is to definitize tasks associated with the fabrication of the third and fourth air vehicles, and long-lead items for the fifth; authorize certain Contractor Acquired Property (CAP); add incremental funding; and revise and update other affected Agreement Articles. Air vehicles number three and four will be built, tested, and delivered to Edwards AFB in support of Phase-III user evaluations. The fifth vehicle will be assembled only upon the exercise of the option to do so. Air-vehicle configuration for numbers three through five is the same as for the two earlier vehicles. Additional activities include building an integrated sensor suite and performing ILS tasks, such as producing technical manuals, training, spares, and addressing reliability and maintainability. These additions are new to the scope of the Agreement.

The Amendment brings the total estimated cost-sharing and performance-fee portion of the Agreement to $230.25 million, total CPFF tasks (covering Contract Line Item Numbers (CLINs) 0001–0008, 0010, 0011) to $112.57 million, and the total award fee pool for CLINs 0007 and 0010 to just over $5 million, for a total Agreement value of $347.94 million. Of this amount, $292.48 million has been obligated.

The Amendment is short (five pages), but the TDD has additional detail. DARPA will conduct four program reviews (Initial Baseline and three Fabrication Reviews) during Phase IIB, and a UFP tracking task has been explicitly required. The TDD states that the UFP will be reduced by improving existing manufacturing documentation, by improving airframe and subsystems producibility, and by using experience from the first two air vehicles during production of the next three. An upgraded and less costly engine, enhanced production tooling, and enhanced manufacturing processes and design modifi-

---

[19]DARPA Agreement No. MDA972-95-3-0013, Amendment No. 0035, 31 March 98.

cations are also mentioned as possible means for reducing UFP. The TDD lists additional NRE and sustaining engineering tasks as an "opportunity to continue development of the Global Hawk system." The TDD provides no schedule information.

SUMMARY OF PHASE II ISSUES AND OUTCOMES

Outcomes to date for the Global Hawk program include:

- *Technical:* The program required unanticipated software development and unexpectedly complex integration tasks. However, the performance objectives of the air vehicle will likely be met, as demonstrated by the three reasonably successful test flights.

- *Schedule:* While the overall length of the ACTD has been maintained, the activity content has been reduced due to slips in interim milestones. First flight was delayed by 14 months (December 1996 to February 1998), the transition to Air Force management slipped by 12 months to October 1998, and Phase III has been compressed from 24 to 15 months. In practice, however, Phase III includes only 12 months of user-evaluation testing because of the need to complete the delayed Phase II flight test program.

- *Cost:* Developmental cost growth for the original Phase II effort has been approximately $70 million, or about 44 percent. UFP is likely to be breached by $2–$3 million, assuming that the original underlying assumptions hold.

- *Quantity:* The Phase III procurement has been reduced from eight air vehicles to a maximum of three; the total quantity has been reduced from a planned ten to a maximum of five. Given fabrication times and the number of systems currently under the Agreement, a maximum of only five air vehicles will be available to support the user evaluation.

- *Other:* Contractor management has changed three times during Phase II, an indicator of a problem program. In contrast, government program management has changed only once, and has been consistent throughout Phase II.

The program became a single-source, best-effort project after funding limitations imposed by Congress during Phase I led to a decision to select a single contractor for Phase II. The lack of competition contributed to subsequent problems. Competition was intended to provide needed discipline because of the accelerated schedule, tight cost caps, and limited government oversight and direction.

While every program experiences unexpected problems, the deviations from plan in this program apparently resulted from several broad and overlapping factors.

- Both the JPO and the contractor underestimated the development task. This shortcoming began in the early origins of the HAE UAV program, when a schedule and budget were prepared before the private sector had performed any serious engineering studies. The projections were not updated on the basis of Phase I data, and now appear to have been optimistic.

- The contractor, TRA, lacked in-house experience in software development and system-integration management for the development phase of projects of this size and complexity, and failed to recognize the importance of this limitation until well into Phase II.

- The integration of COTS equipment into a complex system—a task that few have significant experience with—was more difficult and time-consuming than anticipated.

- In the early phases, most managers viewed the flight vehicle, especially the wing, as a major engineering challenge and risk. They concentrated on those elements, giving short shrift to software development and overall system integration. Perhaps because of that concentration, basic flight-vehicle development progressed largely on schedule. Deemphasized parts of the program—especially software development and completion of an SIL for demonstrating both flight-level and system-level software—lagged far behind schedule and delayed the start of flight testing.

• None of the program structure or management processes specified in Military Standards or other DoD guidance[20] was mandated for this program, consistent with the attempt to conduct
  the project without traditional management and oversight procedures. TRA management took advantage of this freedom and
  eliminated many of the rigorous controls typically required, but
  did not implement procedures of their own to monitor processes. The resulting lack of management rigor and planning in-
  evitably contributed to the evolution of problems.

The effects of these factors on the program were exacerbated by the
constraints on government management intervention built into the
agreement.

One exceptional aspect of the Global Hawk program is that outside
influences contributed little to the schedule slip and cost growth the
program experienced. The Phase I budget cut that resulted in the
loss of competition during Phase II probably affected the process
negatively, although objective measurement would be difficult.
High-level government insistence on maintaining the UFP and TRA's
reluctance to drop functionality likely also contributed to the problems.

This combination of conditions is virtually unprecedented in a major
system-acquisition program. Prior studies show that typical
weapon-system development programs have exceeded their scheduled EMD duration by about one-third, and that most of that excess
can be attributed to guidance, funding changes, and other factors
imposed by outside entities.[21] The relative stability enjoyed by the
Global Hawk program can be traced in part to the special acquisition
management structure used in the program, which prevented significant requirement creep, and high-level sustained support from all
relevant participants, such as DARPA, Congress, and users.

---

[20]See Defense Systems Management College, *Systems Engineering Management Guide*, December 1986. See especially MIL-STD 1521.

[21]See, for example, Drezner and Smith, *An Analysis of Weapon System Acquisition Schedules*, RAND R-3937-ACQ, 1990; *The Affordable Acquisition Approach Study*, Air Force Systems Command, 1983.

Chapter Four

DARKSTAR

This chapter describes the experiences of the DarkStar (Tier III–) program through August 1998. The program has pioneered acquisition innovation; it was the first approved ACTD, and was awarded the first Section 845 Agreement. The DarkStar program was managed by the same JPO as the Global Hawk, and enjoyed the same acquisition flexibilities, but key differences between their program implementations provide interesting comparisons:

- The DarkStar was initiated as a sole-source procurement based on prior work on an analogous (but larger) system, whereas the Global Hawk went through a competitive phase.

- The DarkStar began as a Special Access program, and the requirement for low observability ("stealth") involved a continuing design penalty that the Global Hawk does not share.

- Although both DarkStar and Global Hawk used CPIF Agreements in Phase II, the details of implementation differ. Even the streamlined Global Hawk Agreement looks wordy compared with the terse DarkStar version.

- JPO oversight of the DarkStar was initially austere, even by DARPA standards, with the focus of JPO attention being on the time-consuming Tier II+ competition.

To an extent, the differences in events and outcomes are linked to these program differences.

GENESIS

As discussed above, in January 1990 the JROC approved an MNS for "Long Endurance Reconnaissance, Surveillance and Target Acquisition (RSTA) Capability." In July 1993, the JROC endorsed a three-tier approach to acquiring an "endurance" capability. The third tier was to be the most capable: a stealthy, high-altitude, long-endurance UAV. However, the DSB "summer study" of July 1993 concluded that "existing potential" (*sic*) Tier III programs were either unable to satisfy the MNS or were too expensive. Accordingly, a parallel approach proposed by DARO was adopted: a high-low force mix of complementary systems, together known as HAE UAV. We discussed the Tier II+ previously. In contrast, the companion Tier III– emphasized survivability through very low observability (VLO); was to be a much smaller vehicle; and would be reserved for selected missions (an estimated 20 percent of all HAE UAV missions).

Because Tier III was rejected on the grounds of cost, the relative affordability of the dual-track approach became persuasive. Accordingly, acquisition streamlining and the UFP limits became integral to both DarkStar and Global Hawk.[1]

Lockheed and Boeing proposed a lower-cost design to meet DARO's proposed Tier III– requirement. On 20 June 1994, the Lockheed/Boeing team was awarded a sole-source DARPA Agreement to design, build, and test its proposed Tier III– system. Elements of the program were designated Special Access, and details were withheld from the public until the roll-out almost a year later (1 June 1995).[2]

TIER III– AGREEMENT

The initial Tier III– Agreement between DARPA and the Lockheed Advanced Development Company (LADC), dated 20 June 1994, was 18 pages long.[3] It provided for the design, manufacture, and testing of two proof-of-concept air vehicles, one radar sensor, one EO sensor, and one Launch, Control, and Recovery Station (LCRS), along

---

[1]"HAE UAV Program Briefing for JROC," 8 November 1994.

[2]Specifically, the air vehicle's shape was designated Special Access.

[3]LADC was renamed the Lockheed Martin Skunk Works (LMSW) at a later date.

with required data-link equipment.[4]   Only one of the air vehicles was to be flight-tested under this agreement.   DARPA specified a 21-month program: 13 months to delivery of the hardware, three months of check-out, and five months of flight testing.   However, DARPA gave LADC the option to pursue a 15-month program schedule "at its discretion."

LMSW believed that, from a technical standpoint, the DarkStar was low-risk.   The contractor believed that its prior work provided an understanding of LO surveillance UAVs, and the technology did not seem radical.   The elements of the DarkStar represented well-trodden areas: digital flight controls, aerodynamics of flying wings, LO shaping and treatments, and data links.   The contractor suspected that the LO K_u-band antenna might be the greatest area of risk, considering it represented about $1 million of the UFP.

The procurement environment in 1994 encouraged this perception of low risk.   LMSW was experiencing layoffs as a result of the military drawdown, and new programs were scarce.   The firm was proud of its reputation as the premier contractor for small, technologically advanced special programs, and saw every incentive to take a "can-do" approach with the Tier III– program.

The agreement specified that LADC and Boeing would collaborate on the program by way of a separate Teaming Agreement between the two companies.   This agreement arranged a virtually equal workshare; LADC would be responsible for aircraft fuselage with systems, final assembly, and testing, and Boeing would provide wings, autonomous controls, and avionics integration.

The total estimated costs were $115.7 million; with fixed and incentive fees added, the total price to DARPA became $124.9 million.   For costs beyond $115.7 million, Lockheed and DARPA agreed to a 50/50 cost-share of overrun.   Figure 4.1 below is a representation of the initial Tier III– cost and fee structure.[5]   As cost decreased (moving left along the horizontal axis), fees increased.   Cost plus fee is the cost to the government.   To the degree that an abbreviated program is less

---

[4]The second flight vehicle was to be available as a back-up for the flight test program.

[5]Tier III– Agreement with LADC dated 20 June 1994, simplification of figure on p. 8.

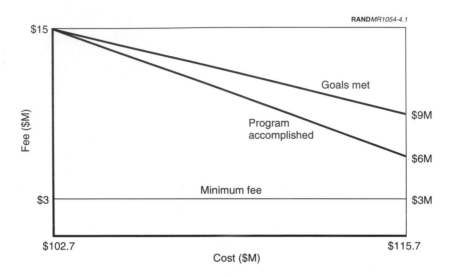

Figure 4.1—Tier III– Initial Cost/Fee Structure

costly, the contractor had clear incentive to pursue the 15-month schedule. Note also that the marginal benefit to the contractor of meeting the more rigorous "program goals" decreased as overall program cost decreased, strengthening the incentive.[6]

The Statement of Work (SOW) included in the agreement specified two goals:

–   To develop a . . . system . . . utilizing the ACTD philosophy, and

–   To develop the proof of concept prototype program so that a cost-effective transition to the operational prototype is achievable.

Subsequent discussion in the SOW addressed the cost side of "cost effective": the $10-million UFP cap, as with the Tier II+, was defined as the average price of aircraft numbers 11–20 in FY94 dollars. However, there was no discussion of a mechanism for transition to

_____

[6]The document referenced for this discussion and Figure 4.1 has limited distribution and was not made available to RAND. Lockheed and the HAE UAV JPO both denied that the gap between "goals" and "accomplishment" criteria was a significant incentive.

production, any criteria associated therewith, or expected changes to the design of the proof-of-concept vehicle. The agreement covered the program only through the end of Phase II (flight test); an amendment extending the program would have been necessary for Phase III (user demonstrations).[7]

Schedule

Figure 4.2 shows an original schedule for the entire planned Tier III– program, including user demonstrations.[8] Note that the period of "limited field demonstrations" designated as part of Phase III was later designated as "Phase IIB." This interim phase was necessary because the Tier II+ schedule lagged behind the Tier III– at that point, and the Phase III user demonstrations required the availability of both systems. However, subsequent Tier III– schedule slippage

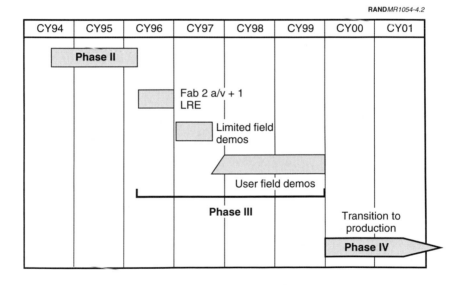

RAND*MR1054-4.2*

Figure 4.2—Tier III– Schedule, Original Program

---

[7]The JPO did not consider this to be an oversight. Rather, the office had not defined those phases and activities well enough to include them in the agreement.

[8]This schedule is derived from one included in an HAE UAV Program Briefing to the JROC, 8 November 1994.

eliminated the phasing problem. Also note that prior work on an analogous (but much larger) system was seen as obviating the need for a Phase I.

Figure 4.3 presents an early Tier III– schedule, covering only Phase II, the part of the program under the DARPA Agreement.[9] Here, Phase II is planned to take 21 months. Note the 16-month period between contract award and first flight, followed by five months of flight tests. The Lockheed push for a 15-month rather than a 21-month Phase II is not reflected; even so, the aggressiveness of the baseline schedule is apparent.

Program Management

The Tier III– was subject to the typical ACTD oversight process, with reporting by the Tier III– Program Manager (PM) up through the HAE

RAND*MR1054-4.3*

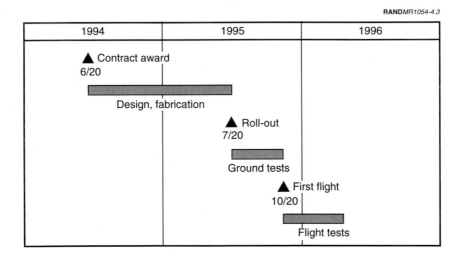

Figure 4.3—Tier III– Phase II Schedule, Original 21-Month Program

---

[9]This schedule is derived from one presented by the JPO Tier III– Program Manager at the Association of Unmanned Vehicle Systems (AUVS) Conference held in Washington, D.C., on 11 July 1995. It represents the original schedule, not the schedule then current.

UAV JPO PM, through the DARPA Director, and then to the DUSD (AT).[10]

Within the JPO, the Tier III– PM drew on expertise available from Tier II+ program personnel. However, the JPO presence at contractor events was limited compared with the equivalent Tier II+ events, with as few as six government staff attending program reviews.[11] It is important to note that in the early days, the Tier II+ side of the JPO was overburdened by its Phase I competition, although it is not known how much this affected the availability of personnel for Tier III– tasks.

Funding

DARPA and DARO shared Tier III– funding responsibility. Table 4.1 presents the originally proposed Tier III– funding plan (refer also to Table 1.2) that would have provided for the delivery of two air vehicles.[12] However, this funding did not cover certain necessary elements of the Tier III– system, which were included in the Tier II+ budget. These included the addition of Tier III– functionality into the CGS; miscellaneous support, studies, and government-furnished equipment (GFE); and all Phase III demonstration costs. Thus the estimated cost of Tier III– was somewhat higher than shown here.

Table 4.1

Original Tier III– Program Funding Plan
(in then-year $M)

FY94	FY95	FY96	FY97	FY98	FY99	FY00	FY01	Total
30	62	50	45	15	15	TBD	TBD	217

RAND MR1054-T.4.1

---

[10]This process is more fully described in Sommer et al., RAND MR-809-DARPA, 1997.

[11]Conversation with Harry Berman, HAE UAV JPO Tier III– PM, 13 November 1996.

[12]"HAE UAV Program Briefing" to the JROC, 8 November 1994.

By July 1995, the estimated total for Tier III– had risen from $217 to $226 million.[13] Since then, program disruptions, some added scope of effort, and an increase in the number of vehicles (four aircraft are now planned as part of the ACTD) have raised the estimated total to $333.7 million, as shown in Table 4.2.[14] Note that the Tier III– items previously included in the Tier II+ line (as above) were moved back into the Tier III– line, making direct comparison between the original and more recent budget lines difficult.

SYSTEM DESCRIPTION

On 1 June 1995, almost two months ahead of the nominal 21-month schedule, and 11 months after the agreement was awarded, the DarkStar air vehicle rolled out. This event signaled the relaxation of the Special Access controls on the program.

The DarkStar was revealed to be a flying-wing design. Selected specifications are listed in Table 4.3.[15] The Tier III– system consists of a CGS in addition to the DarkStar air vehicle. The initial LCRS and associated Processing and Display System (PDS) will eventually be superseded by the CGS, as discussed in Chapter Five.

Table 4.2

**Tier III– Program Funding History
(in then-year $M)**

FY94	FY95	FY96	FY97	FY98	FY99	FY01	Total
42	57	86	65	44	34	6	334

RAND MR1054-T.4.2

---

[13]Presentation by HAE UAV JPO PM at the AUVS Conference, 11 July 1995.

[14]Data provided by Patrick Bailey, HAE UAV JPO, January 1997.

[15]"HAE UAV Program Briefing" to the JROC, 8 November 1994; Tier III– Independent Review Team briefing, 23 September 1996; DarkStar Flight Readiness Review Working Group briefing, 15–16 August 1995.

The accepted 1/200 attrition rate per sortie, the limited design life, the lack of all-weather features, and the constrained flight envelope all point to the program being exactly what it professed to be: a proof-of-concept demonstration to be achieved efficiently and expeditiously. The limited demonstration objective was to show that a loitering, low-observable UAV can carry sensors over the battlefield. It was acceptable for the aircraft to be "attritable" (subject to high attrition).[16] A senior government official said early on, ". . . after fifty hours, bury them." The contractor was to demonstrate sensor functionality prior to the completion of airworthiness flight testing.[17] Whether the initial DarkStar program plan would have resulted in a vehicle robust enough for military utility is an open question, inasmuch as "military utility" has no firm definition. For the unique capabilities offered by the "silver bullet" DarkStar, *any* level of attri-

Table 4.3

DarkStar Specifications

Performance Objective	Value
Design mission altitude	45,000 ft
Time on station at altitude	8 hrs @ 500 nmi
Max gross weight	8600 lbs
Installed payload	1000 lbs
Factor of safety (structural)	1.25
Normal load limit (structural)	+2.0, 0.0
Max roll angle	35 degrees
Airspeed limits (at max wt)	117–140 kts (equivalent)
Takeoff and landing speed	120 kts (at max weight)
Max yaw angle	15 degrees
De-ice	None
Anti-ice	Air data sensors
Design life	300 hrs/50 cycles/3 years
Probability of safe landing per sortie	0.995
Span	69 ft
Length	15 ft

[16]"Attritable" is no longer acceptable. HAE UAV ACTD Users' Conference, NASA Dryden, 25 February 1998.

[17]Harry Berman, "DarkStar—High Altitude Endurance UAV," paper presented at Unmanned Vehicles '97 Conference and Exhibition, Paris, France, 12–13 June 1997, p. 11.

tion might possibly be acceptable to an operational planner.[18] The key issue is whether the initial government focus on an attritable system, and the correspondingly limited available funding, was appropriate.

The rationale behind the design of the DarkStar air vehicle casts light on the program's initial assumptions regarding risk.

The requirement for low observability drove the unusual design of the DarkStar, which was primarily achieved through shaping. Shaping is a hidden cost of stealth design, because it fundamentally determines all other vehicle characteristics—unlike radar-absorptive material, the weight and cost impact of which is simply additive. In DarkStar's case, the combination of the loitering mission and stealth capability led inexorably to the choice of an unswept flying wing. The sensor payload was then put into a semi-circular, sharp-edged fuselage that gave an acceptably low omni-azimuthal radar return. The DarkStar planform is depicted in Figure 4.4.

Figure 4.4—DarkStar Planform

---

[18]Only the completion of the user-demonstration phase can answer this question. If the scope of the user demonstration phase is inadequate to determine "military utility," the question may be unanswerable.

The resulting design was, by necessity, compromised aerodynamically. This type of compromise was not new to LADC, by virtue of their experience with the F-117 stealth fighter. However, a flying wing with no significant sweep or tail has an inherent problem in generating pitch forces in response to atmospheric disturbances or takeoff requirements. The geometry places unusual demands on the aircraft flight-control system, which must respond rapidly to keep the wing within a low-range angle of attack.[19] The problem is amplified on the ground if the vehicle has a rear-set main-landing-gear configuration, such as the DarkStar has.

A lightly loaded, high aspect-ratio wing, such as that found on the DarkStar (necessary for endurance), can lead to additional problems. Gust response is magnified, ground-effect interactions become larger (a concern during landing and takeoff), and aeroelastic effects (wing bending) complicate the design of the flight-control system.[20]

The DarkStar designers were aware of these aerodynamic complications, but because of the intent of the program to get the sensors in the air as soon as possible, and the attritable design philosophy, the program forged ahead.

EVENTS

LADC clearly wanted to meet the 15-month program schedule from the start, as was their option under the agreement. The agreement provided incentives for both cost and performance, but LADC chose to make decisions on the basis of cost early in the program.[21]

The DarkStar configuration was frozen 11 days after the award of the agreement. Shortly thereafter, Boeing wind-tunnel tests revealed

---

[19]As of 1995, the DarkStar was limited to 3–5 degrees of body angle of attack, depending on speed, this being the source of the 35-degree angle of bank limitation in Table 4.2. DarkStar Flight Readiness Review Working Group briefing, 15–16 August 1995.

[20]The DarkStar is essentially a wheeled wing-in ground-effect machine during takeoff and landing, being in deep ground effect (height/span < 1/20). The technology base for this type of craft is sparse; specialized wind-tunnel facilities or actual flight testing are required to validate analytical simulations—steps that were not taken in this program.

[21]Conversation with Harry Berman, HAE UAV JPO Tier III– PM, 13 November 1996.

that the fuselage of the DarkStar was providing unexpected excessive lift and drag at operational angles of attack, because of vortex formation. Under its compressed schedule, LADC addressed the problem by inserting two-foot wing plugs into the wing roots, twisted so as to decrease the body's incidence by two degrees and rebalance lift forces at expected flight angles of attack.[22]

In the six months following roll-out, all parties realized that the first flight would be delayed. A number of problems were cited. LADC discovered that it had underestimated the difficulty of adapting certain existing FCS software to the DarkStar; integration of COTS avionics was unexpectedly difficult; and the ARC-210 radios, planned for wide use in the military, needed modification for the DarkStar.

During this time, no further indications of aerodynamic problems arose. The last Boeing wind-tunnel test took place in February 1995.[23] Unknown to Boeing, however, a significant moment reference error was made during the course of this test, which contributed to a later discrepancy between flight-test data and simulation.[24]

From January through March 1996, LADC conducted taxi tests ranging from speeds of 30 knots to 78 knots. LADC originally intended to conduct taxi tests up to 90 knots, but "growing confidence in analytical models of the aircraft . . . led contractors to suggest that preflight taxi tests could be cut short," and the 90-knot test was never conducted.[25]

On 29 March 1996, the first DarkStar air vehicle had its first flight. Although the flight itself was uneventful, the takeoff run had a dis-

---

[22]Dornheim, Michael, "Mission of Tier 3– Reflected in Design," *Aviation Week & Space Technology,* 19 June 1995, p. 55.

[23]Tier III– Independent Review Team briefing, 23 September 1996.

[24]Tier III– Independent Review Team briefing, "IRT-2," 2–5 December 1996.

[25]Tier III– Independent Review Team briefing, 23 September 1996; Fulghum, David A., "DarkStar First Flight Possible in March," *Aviation Week & Space Technology,* 19 February 1996, p. 53; "DarkStar in Taxi Tests," *Aviation Week & Space Technology,* 26 February 1996, p. 23. The JPO claims that an assessment of the risks of high-speed taxi also affected this decision.

tinct anomaly: the aircraft sped down the runway on its nosewheel, "wheel barrowing" for about 100 yards.[26]

The engineers knew prior to the second flight that there was sufficient disagreement between predicted and actual flight data to suggest that the simulation models were flawed. These engineers, and their IPT leads, told program management that they were not yet ready for the second flight. Management relayed this message to the HAE UAV JPO. However, contractor program management was determined to meet the shortened schedule, and, with support from DARPA's highest level, pushed ahead despite the warnings.

In response to this anomaly, LADC made changes to the FCS software and takeoff technique, and made a second flight attempt on 22 April 1996. The aircraft crashed on takeoff, after a pitch oscillation developed during the high-speed ground run (about 90 knots), causing the vehicle to pitch up uncontrollably just after leaving the ground.[27]

An Accident Investigation Board was promptly formed. The board concluded that:[28]

- The primary cause of the accident was interaction of the ground, landing gear, and air-vehicle mass.

- The air-vehicle control system was unable to damp out the oscillations.

- Insufficient fidelity existed in the simulation.

- The test team lacked situational awareness.

An Independent Review Team (IRT) was established to assist in putting the DarkStar program back on track.[29] The IRT was to assess:

---

[26]Dornheim, Michael A., "DarkStar Makes 'Solo' First Flight," *Aviation Week & Space Technology*, 8 April 1996, pp. 20–22.

[27]Dornheim, Michael A., "DarkStar Destroyed on Second Flight," *Aviation Week & Space Technology*, 29 April 1996, pp. 24–25; "HAE Oversight Briefing," 12 November 1996.

[28]"HAE Oversight Briefing," 12 November 1996.

[29]The nine IRT members included the LADC Vice President of Engineering and a Boeing Systems Test representative.

- contractor understanding of the system

- processes and procedures being used to identify a robust system solution

- proposed software and hardware changes that would lead to a robust system

- the system test and evaluation plan.

The IRT met twice, during 23–27 September 1996 and 2–5 December 1996. The assessment concluded with a briefing to the USD(A&T), Dr. Paul Kaminski. The revised program plan was approved by the USD(A&T), including the planned acquisition of two additional air vehicles to be developed for demonstration purposes.[30] The IRT recommended that LADC:

- improve fidelity of simulations

- improve robustness of FCS software through stress testing

- use a hiking nose gear to prevent oscillations on ground roll (this would keep the aircraft at zero lift until it achieved a higher speed, then would extend the nose gear for rotation and flyaway)

- examine program risk-management methodology

- improve flight-test team situational awareness.

The technical and procedural changes addressing the takeoff problem were judged to be sub-optimal. Flawed landing-gear location was the proximate cause of the crash. While moving the main gear forward would have resolved the immediate takeoff problem, this would have required a major vehicle redesign that would be prohibitively expensive.

The IRT made some interesting observations (that were not included in the final decision briefing to Dr. Kaminski) that pertained to the overall program's philosophy and its environment:

---

[30]"Kaminski Approves Revised DarkStar UAV Program," DoD electronic news feed, 21 January 1997.

- "More ATD than ACTD environment"[31]

- "Schedule pressures cause stress on especially system software and flight test processes"[32]

- "Adequate time was not allowed for evaluation of the first flight data, nor flight validation of the simulation. This, even though there was a sense of serious problems that were not understood . . . The same schedule driven mentality seems to be prevailing at this present time with regard to fixes."[33]

- "As currently configured, the Tier III– project includes technologies that are relatively new, and involves technical risks of various extent. These technologies are: completely autonomous flight operations; active modal suppression system for the first symmetric wing bending mode; integration of the nose gear hiking system into the longitudinal flight control system."[34]

- "Root source, Tier III– robustness problem, is a wing that stalls at AOA=5 degrees while the body continues to lift, causing the shift forward. Aero re-design does not appear feasible due to endurance and RCS issues."[35]

- "My general impression is that the design may be made acceptable for flight testing, but its operational utility may be severely compromised by lack of robustness."[36]

- "Lack of redundancy in the overall system design and limited experience with lightly loaded, high aspect ratio, tailless

---

[31]"General Observations," DarkStar IRT#1 Report, LMSW, 27 September 1996. This refers to Advanced Technology Demonstrations (ATDs). ATDs evaluate only technical performance, not military utility, and are not expected to transition into production.

[32]"General Observations," DarkStar IRT#1 Report, LMSW, 27 September 1996.

[33]"Specific Findings/Recommendations," DarkStar IRT#1 Report, LMSW, 27 September 1996.

[34]"Specific Findings/Recommendations," DarkStar IRT#1 Report, LMSW, 27 September 1996.

[35]"Specific Findings/Recommendations," DarkStar IRT#1 Report, LMSW, 27 September 1996.

[36]"Specific Findings/Recommendations," DarkStar IRT#1 Report, LMSW, 27 September 1996.

configuration results in an interesting test bed, but <u>NOT</u> in a vehicle with safe and routine operational capability."[37]

Both LMSW and JPO downplayed the significance of these observations in conversations with RAND, on the basis that the IRT approached the problem from a manned-aircraft perspective, and did not understand UAVs. We judged this to be evidence of a continuing cultural chasm separating UAV designers from the manned-aircraft mainstream, one that goes to the heart of the issue of UAV risk tolerance or risk aversion. Experts in the field clearly had strong reservations about the inherent robustness of the DarkStar design, notwithstanding the selected and implemented post-crash modifications.

Flight Test Hiatus and Resumption

The DarkStar flight test program was deferred for 26 months, from 22 April 1996 to 29 June 1998, while the JPO and its contractors attempted to redesign the vehicle. System "adjustments" included:[38]

- rewrite of the flight-control system laws

- system simulation changes

- total redesign and remanufacture of the landing gear, and integration of the landing gear into the flight-control system

- takeoff methodology changes

- communications system changes

- changing the crew-training syllabus to focus on emergency procedures.

In March 1998, the second DarkStar air vehicle commenced a series of taxi tests. These taxi tests were conducted up to planned rotation speed, far beyond the 78 knots tested with the crashed air vehicle. Finally, the second DarkStar enjoyed a successful first flight on 29

---

[37]"Specific Findings/Recommendations," DarkStar IRT-2 Report, Seattle, 2–5 December 1996.

[38]Berman, Harry, "DarkStar—High Altitude Endurance UAV," paper presented at Unmanned Vehicles '97 Conference and Exhibition, Paris, France, 12–13 June 1997, p. 10.

June 1998, although the aircraft experienced an anomalous pitch oscillation. Further flight testing continues.[39] The primary disadvantage to the nose-gear hiking scheme apparently is an increased takeoff run, relatively longer at lower gross weights. Whether these limitations will affect the determination of military utility remains to be seen. Takeoff performance may not be stressed during the user demonstration, because almost all flights will operate out of the Edwards AFB flight-test center, where runway lengths are ample.

The program delays and consequent cost overrun brought financial pressure on LADC and Boeing. Amendment 21 to the DarkStar Agreement specified a new incentive-fee structure,[40] depicted in Figure 4.5. In this figure, find the total price to the Government by adding x and y coordinates (i.e., cost plus fee). Thus, at the maximum estimated cost of $183 million for the basic DarkStar SOW (CLIN 0001), and assuming technical performance goals are met, LMSW receives a base fee of $3.1 million and a technical performance fee of $5.6 million, but its cost share of the overrun amounts to $25.6 million. The total fee less cost share is therefore about minus-$17 million. The net price to the government is $183 million minus $17 million, or $166 million. This represents a 33-percent growth in price over the maximum estimate at program initiation ($124.9 million).[41]

The most recent DarkStar Agreement modification available to RAND is Amendment 43, dated 1 July 1998. As of that date, the maximum estimated cost for the basic DarkStar SOW had increased to $205 million. The base fee was still $3.1 million and the technical-performance fee $5.6 million, but contractor cost share of the overrun had increased to $33.7 million. The total-fee-less-cost-share therefore is about minus-$25 million. The net price to the govern-

---

[39]"Second DarkStar UAV Completes Test Flight," DoD News Release No. 327-98, 29 June 1990.

[40]Amendment 21 to MDA972-94-3-0042, 1 July 1996. This amendment provides for bringing the second air vehicle into the flight-test program; incorporates changes resulting from the IRT process; and redefines performance goals and incentive-fee structure.

[41]The plateau in the total-fee-less-cost-share curve is because of the effect of added level of effort following the crash. Only CLIN0001 (basic DarkStar SOW) is considered, to eliminate the effect of other changes in scope.

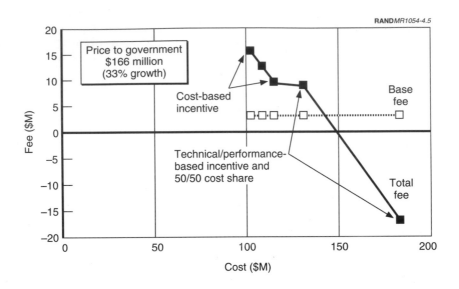

Figure 4.5—Effect of Cost Growth on Contractor Fee As of July 1996

ment is $205 million minus $25 million, or $180 million. This now represents a 44-percent growth in price over the maximum estimate at program initiation ($124.9 million).[42]   See Figure 4.6.

Figure 4.7 presents the information in chronological form. The CLIN 0001 total includes the contractor base and performance fees, and these fees do not offset the cost-share curve. CLIN 0001 represents the scope of the basic DarkStar SOW; the large jump in July 1996 is because of the effect of Amendment 21, as described above. CLINs 0002–0010 cover the introduction of CGS cooperation; T1 SATCOM; NASA Dryden test transportation; LO testing; special studies and O&M assessments; long-lead items for aircraft three, four, and five; additional payloads; and additional NRE items and studies. Some of

---

[42]The base fee was actually paid to the contractor as work progressed. The contractor has not yet earned the technical-performance fee. In this and other DarkStar cost calculations, we have assumed that the technical-performance fee will eventually be paid.

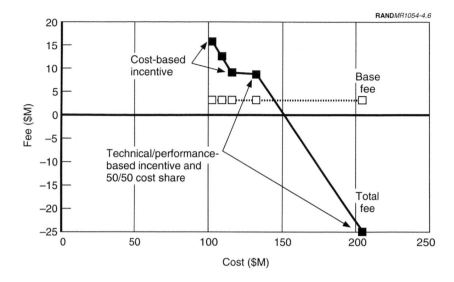

Figure 4.6—Effect of Cost Growth on Contractor Fee As of July 1998

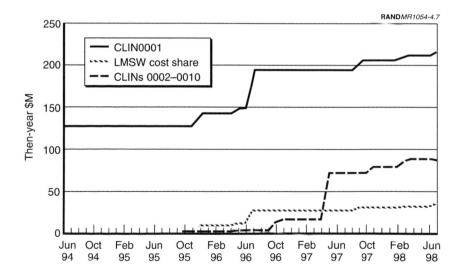

Figure 4.7—Evolution of DarkStar Program Costs

the activities included in these CLINs may relate to completion of the spirit or intent of the original SOW, but a detailed allocation of these costs is not possible. Thus, Figure 4.7 understates the cost growth for the original SOW by an unknown amount.

In 1997, the principals planned to delay the start of user demonstrations by one year, to 1999. Figure 4.8 presents a DarkStar program schedule deriving from that assumption.[43]

Because of the planned delay and increasing congressional concerns about cost growth and schedule delays experienced by the HAE UAV program as a whole, it was determined that the funding required to delay the production decision would be denied. Consequently, the time available for both initial flight testing and for user evaluations has been slashed, as represented in Figure 4.9. The "user field demonstrations" accommodate this reduction by re-designating the

RAND*MR1054-4.8*

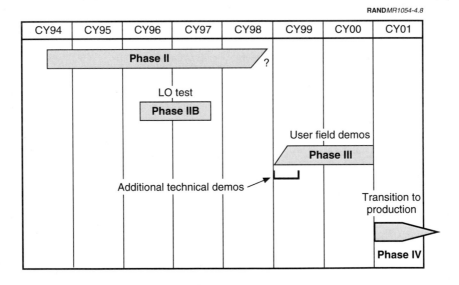

Figure 4.8—Revised DarkStar Schedule As of Mid-1997

---

[43]The additional technical demonstrations during Phase III are referenced in the "HAE UAV ACTD Users Conference Briefing," 7–8 January 1997. The delay in commencement of user demonstrations is noted in the DARPA Tier III– PM letter to RAND, 3 July 1997. The time period for Phase IIB is specified in Amendment 0024 to Agreement MDA972-94-3-0042, 18 October 1996.

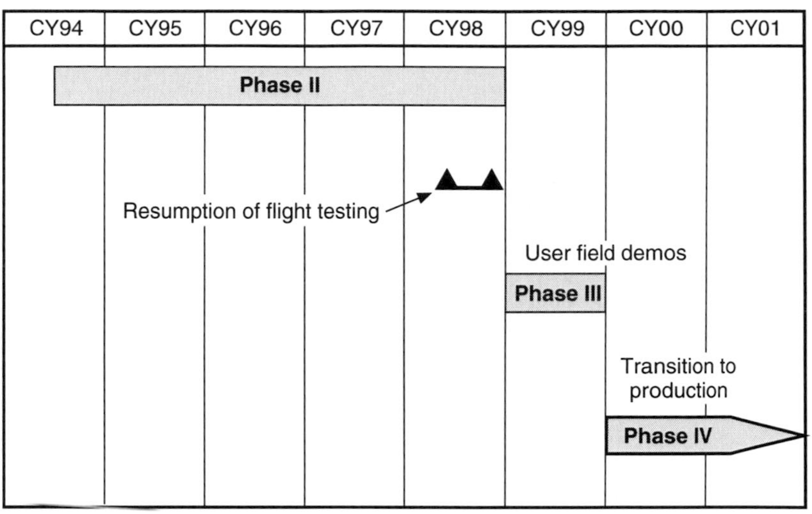

RAND*MR1054-4.9*

CY94	CY95	CY96	CY97	CY98	CY99	CY00	CY01

Phase II

Resumption of flight testing

User field demos

Phase III

Transition to production

Phase IV

Figure 4.9—DarkStar Schedule As of Mid-1998

first quarter as a period of "technical demos," reflecting that flight testing will not be completed in 1998. This reduces the scope of DarkStar participation in user exercises. The real question is whether the users' determination of military utility will be possible in the limited time available. The availability of sufficient information to support a production decision grows increasingly unlikely. For example, all but two of the potential exercises have the UAVs based at the Edwards AFB flight-test center, and airspace access will be negotiated on a case-by-case basis. Because airspace access is a major issue for UAVs in general, and solutions could involve significant changes in UAV control philosophy and avionics-fit, committing to production before those issues are resolved might be premature. Weather is another issue: DarkStar may not have the opportunity to perform in sufficiently diverse conditions to demonstrate its robustness in service use.[44]

_____

[44]HAE UAV ACTD Users Conference, NASA Dryden, 25 February 1998. This discussion is also relevant to Global Hawk.

Finally, Figure 4.10 summarizes the schedule history of the program since its inception. The open triangle represents the de facto initial target schedule (15-month program). Note that the first major difficulties in the program did not become apparent until 12–18 months after contract award (i.e., until after roll-out). Even after the crash, schedule estimates were unduly optimistic.

CONCLUSIONS

Although it is too early to reach definitive judgment on lessons learned from this program, planners of innovative acquisition programs should already find points for consideration.

First and foremost, the designation of the Tier III– program as an ACTD was a critical decision. Notwithstanding LMSW and JPO opinions to the contrary, the aerodynamic configuration of the air vehicle represented an immature technology. Likewise, the software revisions to improve robustness represent novel technologies, as the IRT pointed out; risks associated with the configuration were poorly understood. Flying the aircraft in the smallest flight envelope pre-

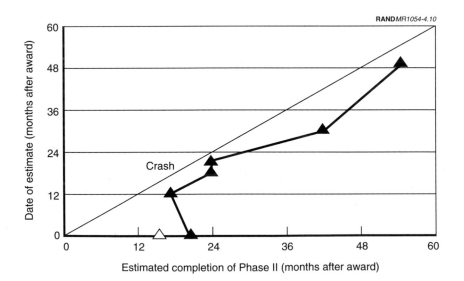

Figure 4.10—DarkStar Schedule History

dicted for sensor demonstration requirements to avoid short-term risks[45] all but guaranteed a user failure downstream, because users then might underappreciate the system's fragility.

Another issue relating to "mature technology" is systems integration. The ACTD Master Plan states that, "by limiting consideration to mature technology, the ACTD avoids the time and risks associated with technology development, concentrating instead on the integration, and demonstration activities."[46] One could assert that no technology is mature in the aggregate if it requires extensive integration of subsidiary technologies, even if those component technologies are themselves mature. Under these assumptions, a system—or "system of systems"—requiring extensive integration should not be an ACTD.[47]

Would these problems have happened in a non-UAV program? Perhaps, but they were doubtless exacerbated because UAVs were involved. In the absence of formal rules and procedures in an acquisition system, the informal guidance of a technical or operational culture can channel efforts into historically productive pathways. However, UAVs are outside traditional aviation culture; thus, the participants are continually learning new lessons. Even under the auspices of two of the world's most capable aerospace companies, a UAV program can become isolated. Given the history of UAV programs, a more risk-averse program strategy probably would have been a better choice.

---

[45]Tier III– Independent Review Team briefing, 23 September 1996.

[46]ACTD Master Plan, August 1996, pp. 1–4.

[47]This is an observation on how ACTD policy affected DarkStar, and not on DarkStar itself.

Chapter Five

COMMON GROUND SEGMENT

During the initial planning for the HAE UAV program in late 1993 and early 1994, it was apparent that a "ground segment" for each system was needed. That segment would perform several functions: control the flight vehicle; receive the sensor data transmitted from the vehicle and transform it into a format suitable for delivery to the users; and provide communications with users, system operators, and others affected by—or benefiting from—the operation of the system. However, the notion of one ground segment that would provide these services to both DarkStar and Global Hawk existed only in preliminary form during the early phases, evolving incrementally over the next two years.

In this chapter we summarize the evolution of the CGS and describe it as it exists today. The discussion includes the ground segments for the Global Hawk and DarkStar, upon which CGS is based.

DESCRIPTION OF ORIGIN AND PURPOSE

The agreement for development and subsequent demonstration of DarkStar was signed in June 1994. In that agreement, the Statement of Work included design, development, and fabrication of an LCRS to provide mission planning, command and control, and communications. The DarkStar ground segment also included a PDS for image processing and dissemination.[1] No one mentioned integrating the

---

[1] The PDS was later called the DarkStar Processing Element (DPE), but was functionally the same.

LCRS or PDS with the Tier II+ system; indeed, such mention would have been premature because the Tier II+ system was little more than a concept at that time, and would not be defined for another nine months.

The next step in the evolution of the overall HAE UAV program was the June 1994 release of the Tier II+ Phase I solicitation. That solicitation required the contractor to design a ground segment for Tier II+ that would provide command and control, mission planning, data processing, and data dissemination. Again, there was no mention of integrating the Tier II+ and III– ground stations into a common system. An admonition to "make every attempt to design a system which is compatible with existing systems" referred primarily to linking the UAV system with existing communications, imagery-handling, and exploitation systems so that the UAV could fit easily into the existing infrastructure without requiring expansion or modification of that infrastructure. However, JPO management was already thinking of a ground segment that would be common to both UAV systems by that time; one of the six IPT teams within the JPO was devoted to that task in organization charts dated 1994. There was external pressure on the JPO to consolidate the Tier II+ and Tier III– ground stations, notably from ADM Owens, then Vice Chairman of the Joint Chiefs of Staff.

The agreement between TRA and DARPA for the Phase I study, dated November 1994, calls for development of a ground station to support the TRA design, but does not mention a common system nor coordinating with the Lockheed design. However, as the JPO reviewed the completed Phase I design studies, the idea of integrating ground segments between the two systems began to form. The solicitation for Phase II of Tier II+, dated 15 February 1995, included preliminary design activity for integration of Tier III– functionality.[2]

The desire to combine at least some elements of the ground segments for the two systems outpaced the strategy and resources for doing so. Likewise, there was no apparent analysis undertaken of the operational benefits of a common ground segment. An amendment to the agreement for Phase II of Global Hawk, dated July 1995, in-

---

[2]The Tier II+ ground segment was also supposed to be sized to accommodate Tier III– functionality.

cluded "support for the Tier III Minus Ground Segment–UAV interface definition," along with the design, development, fabrication, and testing of the Tier II+ Ground Segment.[3] The Global Hawk Phase II work statement did not impose the obligation of creating a common ground segment, but of examining the integration of the ground segments of the two systems.

Within the Global Hawk program, the ground segment was the responsibility of one of TRA's main team members: E-Systems, Inc. In addition to the main task of designing and building a ground system to support the Global Hawk flight tests and operational demonstrations, E-Systems studied the concept of a common ground segment. By October 1995 the firm had created an integrated set of requirements that such a system would have to meet, and by year's end had offered a common ground segment design to the JPO. Unfortunately, the task of integrating all the imagery and communications elements in such a system, as well as the daunting task of developing and proving the software, would require resources beyond those available at that time.

In January 1996 the JPO reached an undefinitized agreement with E-Systems that authorized initial work on two critical segments of the system; mission planning and communications. Meanwhile, TRA was struggling with the development of a system that would interface with, and affect, Lockheed's work on the DarkStar LCRS. Some time in mid-1996 the JPO realized that progress would be enhanced if TRA was relieved of management responsibility for development and integration of the common ground system, and began negotiations with E-Systems toward that end. In March 1997 DARPA signed a definitized agreement with E-Systems authorizing the contractor to "design, develop, integrate, and test modifications to the Global Hawk Ground Segment to accommodate operations of the DarkStar UAV. The team will modify one Global Hawk Ground Segment and demonstrate system performance." The stated objective of the agreement was to achieve DarkStar operations with the CGS and demonstrate simultaneous operations with Global Hawk. The agreement indicated that the CGS should include integrated mission planning, command and control for both air vehicles, data-link

---

[3]The Global Hawk ground segment is composed of an LRE and an MCE.

communications, data processing, and integrated imagery dissemination.

According to the DARO UAV Annual report, the CGS is intended to "provide digital, high quality, near real-time imagery to warfighters and users at various command levels." The CGS is intended to control up to three air vehicles in any combination through line-of-sight (LOS) data link and SATCOM relay.[4] There is no fixed design price, but an informal $20-million unit-cost goal was established. Substantial use of OTS software and hardware was planned.

BASELINE PLAN

As discussed above, the original Tier II+ Agreement refers to the TRA ground segment: E-Systems is a subcontractor to TRA and is responsible for the Global Hawk ground segment (LRE and MCE). A modification to this agreement included initial consideration of incorporating Tier III– functionality into the Tier II+ ground segment.

DARPA and E-Systems signed the initial agreement for CGS on 24 January 1996.[5] It was an open-ended, undefinitized letter contract with a $1-million ceiling that allowed E-Systems to begin mission-planning and interface-definition software development. It referenced the 25 November 1995 Technical Proposal,[6] but did not define specific deliverables or the incentive-fee structure. The contract term was through March 1996 with the expectation that the parties would negotiate a definitized agreement by the end of February 1996. The agreement was amended 10 times to extend the period of performance, increase the not-to-exceed cap, and add activities.

The functional baseline for the CGS program is Amendment 11, dated 31 March 1997. This amendment tightened and made definite the initial agreement awarded in January 1996. It estimated the cost of the project at $25 million, designated Raytheon E-Systems as prime contractor, and assigned E-Systems the integration of the

---

[4]Defense Airborne Reconnaissance Office, *UAV Annual Report,* FY1996 and FY1997.

[5]Agreement MDA972-96-C-0006.

[6]"HAE UAV Common Ground Segment Proposal: Technical Approach," 25 November 1995, Raytheon E-Systems Falls Church Division.

original LCRS and PDS functionality from DarkStar into the Global Hawk ground segment. E-Systems would accomplish this by adding hardware and software, in the form of modification kits, to the Global Hawk MCE and LRE to incorporate DarkStar (air vehicle number 2 configuration) mission-planning and control functions. E-Systems would ship the Tier III– upgrade package to the Tier II+ ground segment as a kit to be installed onsite wherever the LRE and MRE were located at that time—most likely at Edwards AFB, in support of flight tests.

A CGS 2 Agreement, to be proposed formally in early October 1997, was to be the first new "integrated"-design ground segment. In June 1997, an amendment to the agreement authorized $2.6M in long-lead funding for the integrated LRE. This amount was increased to $17 million in March 1998 to procure the second CGS. Between the three activities—original Global Hawk ground segment, CGS 1 for the modification kits, and CGS 2 for the new design common ground station—E-Systems would produce two LREs and two MCEs. CGS 2 was to support the DarkStar configuration for the third and fourth air vehicles, as well as Global Hawk. Under CGS 1 and 2, E-Systems sub-contracts with both TRA and LMSW, as well as Boeing Military Aircraft Division, Northrop Grumman Electronic Sensors and Systems Division, Lockheed Martin Wideband Systems, and GDE Systems Incorporated.

CGS 1 Agreement (Amendment 0011)

The objective is to modify the Global Hawk ground segment to accommodate the operations of DarkStar. This is essentially a "best effort" contract with respect to E-Systems' performance; neither funding nor fee is directly tied to specific system-performance parameters.

The agreement defines a combination of CPIF and CPFF work. Item 0001—the basic design, development, and demonstration of the CGS—is CPIF, with a total projected cost as follows:

Target cost	$23,930,593
Target fee	$1,895,492
CPIF	$25,826,085

The target fee was 8 percent of the target cost, bounded by a zero minimum fee and a maximum fee of 10 percent. In addition, certain support tasks were to be performed under CPFF with total projected costs as follows:

0002	Direct downlink demonstration kit	$1,200,000
0003	ACCM concept validation	$48,341
0004	CIGSS support	$9,000
0005	R&M of GFE	$200,000 (NTE)
	Total Estimated Cost	$1,375,040
	Total Fee	$82,301
	Total Cost Plus Fee	$1,457,341

The sum of those two categories leads to a total projected cost to the government of $27,283,426. Of that, the government paid $20,457,341 to cover work accomplished during 1996 and the future work needed to complete the project.

The agreement also included a plan for an additional award fee on top of the target fee applying to CLIN 0001. A total pool of $1 million is available for this award fee, with specific awards to be evaluated at designated program milestones and dates: design review, 31 March 98; CGS/DarkStar flight test, six months after completion of DarkStar flight test; and agreement completion.

We should note that these costs are in addition to costs allocated to developing individual ground stations for the two UAV programs. Costs of the DarkStar LCRS are not reported separately, but cost of the Global Hawk ground support system was $24.2 million through the end of FY95.

The agreement was vague about schedule, which is almost inevitable. To perform the task, E-Systems had to use both information and simulation facilities provided by the developers of DarkStar and Global Hawk, and demonstration of the system depended on availability of operating flight vehicles from both programs. The agreement provided only two estimates of overall schedule. Article 4 (Term) stated that the agreement commenced on 14 January 1996 (date of the first agreement by DARPA and E-Systems for development of a CGS) and "continues for the duration of the effort

... which is approximately 38 months." That would put the end of the CGS development activity around March 1999—about 15 months later than the estimated completion of the CGS shown on the mid-1995 program schedule (see Figure 1.2) but consistent with the overall schedule of the two flight vehicle programs it supported.

The baseline schedule for CGS was tied to the Global Hawk schedule. The IDR, DRR, and FDR, as well as system tests, were essentially the same as Global Hawk, according to the August 1996 Technical Baseline and Program Plan.

The TDD identified a single "deliverable": one modified Global Hawk ground segment. The TDD also defined each CGS team member's role and specified the responsibilities of each CGS industry team member, including hardware and software development. The interfaces between contractors were defined and managed by Interface Definition Documents. An Engineering Change Notice process was established to review and manage changes after drawing approval.

TPMs were established for tracking actual versus planned performance, to provide early identification of problems, and as a tool for assessing the impact of changes. The TPMs were to be part of the monthly status reviews, as well as the formal design review. The program had 12 software TPMs, 10 hardware TPMs, 11 segment/element level test TPMs, and 5 performance TPMs.

Four IPTs managed the CGS: program management, ground segment, systems engineering, and test and support. Either JPO staff or SETA contractors representing the JPO were explicitly part of the IPT membership, with the same responsibilities as E-Systems personnel.

The agreement resembled the other government-industry arrangements in the HAE UAV program. The agreement allowed internal audit and cost-accounting processes. As in the Global Hawk and DarkStar Agreements, either party could terminate the contract with no penalty. There were 17 FAR Clauses called out under Article 21, in addition to a data-rights FAR clause. The contractor retained patent rights.

Several assumptions underlay the agreement. The cost estimate was based on five months of integration and testing at Edwards AFB. All documentation was to be provided in contractor format. DarkStar air vehicle #2 was the configuration baseline for that system. O&M costs were handled separately. DARPA would conduct only one formal design review; all other reviews defined in the IMP were to be "peer level." Both Global Hawk and DarkStar were to have flown with on-board sensors by 1 April 1998.[7] Software integration labs for both DarkStar and Global Hawk were to be available throughout the period of performance.

CGS 2 Amendment

The agreement was amended in March 1998 to authorize and fund the design, fabrication, and test of the second, more-integrated CGS. As mentioned above, this was priced at $17 million. The amendment also added approximately $9 million to the CGS 1 to accommodate revisions accruing from changes to the DarkStar and Global Hawk air vehicles. The total award fee was increased to $1.7 million. Other agreement provisions remained unchanged.

EVENTS TO DATE

The ground segments for the two air vehicles have not driven technical, cost, or schedule outcomes in the HAE UAV program. For the most part, the ground-segment efforts have proceeded on track. As noted in Chapter Three, initial tension between contractors regarding the CGS has not been fully resolved. However, the government-industry relationship (the ground-segment IPT, led by E-Systems) experienced none of the tension that the air vehicle IPTs encountered.

As mentioned previously, LMSW and TRA were effectively in competition in terms of the force-mix decision pending at the end of the ACTD program. Thus, the two contractors had no incentive to cooperate in developing a CGS. As a consequence, the two air vehicles and their systems were developed essentially independently. Additionally, that information E-Systems required for CGS development was not always available when needed. The separate agreement

---

[7]As discussed in Chapters Three and Four, this milestone was not achieved.

between DARPA and E-Systems was intended in part to resolve this issue; however, it placed the government in the role of systems integrator. The JPO and the contractors agree that the government historically has not fulfilled this role well. Additionally, in the HAE UAV acquisition strategy, most design and management responsibility rests with the contractors, making a government integration role even more problematic.

The January 1996 letter agreement allowed E-Systems to begin work on interface-definition and mission-planning software development. In the original plan for the CGS, items requiring modification in order to integrate DarkStar functionality into the Global Hawk ground segment included:[8]

- LRE/MCE command and control workstation and software

- LRE/MCE mission-planning, communications, and image-quality-control software

- LRE/MCE UAVIP and data-link hardware and software

- LRE/MCE server (data processing hub)

- LRE/MCE SGI Challenge XL and Mercury EO/IR data-processing suites

- LRE differential GPS.

This is a substantial portion of the subsystems comprising the ground segment. The Tier II Plus LRE/MCE command and control, telemetry, and sensor-data platform interfaces are functionally redundant to the Tier III Minus because of the independent air-vehicle development.

As work proceeded, a truly common ground segment was acknowledged as too costly, particularly in terms of software development. The estimated cost of the November 1995 proposal exceeded the available budget. Differences in mission equipment between DarkStar and Global Hawk that provide different types of data constrained commonality. Furthermore, LMSW did not provide certain

---

[8]Raytheon E-Systems, "Tier III– Integration Proposal: Technical Approach," 22 November 1995.

key interface data in a timely fashion. As of August 1996, the CGS plan acknowledged that common mission-planning goals would be only partially met, common dissemination goals could be fully met, and integrated communications, command and control and image processing would not be met.[9] While not all ground-segment functions will be integrated in the current configuration, a roadmap exists to fully integrate all functionality when funds become available.

The Global Hawk LRE was planned for delivery on 1 October 1996 in support of Global Hawk's first flight, then planned for 17 December 1996. Various events caused a delay: actual delivery of the LRE was December 2, only two months behind the original schedule and one of the first significant milestones in the Tier II+ program. After delivery, the LRE operated around-the-clock at TRA in support of the Tier II+ program for 10 months with no catastrophic failures.

The Global Hawk MCE left E-Systems for delivery to TRA on 30 September 1997, 11 months behind the original schedule; the original plan was for November 1996. From a contractual standpoint the MCE was delivered on schedule, although 11 months behind the original schedule.[10] However, given other program delays, the MCE was not on the critical path.

Potential technical problems began to appear in mid-1998. At the end of July, indications surfaced that the LRE may not be able to handle both the Global Hawk and the DarkStar simultaneously. E-Systems has been addressing this problem. Also, the MCE's ability to disseminate information (imagery) to exploitation cells was a growing concern. The JPO considered this a critical problem, because the utility of the HAE UAV systems depends on dissemination of imagery in common format.

The informal CGS unit price goal of $20 million in FY95 dollars is still in place. The breakout is $4 million for the LRE and $16 million for the MCE. The $50 million funding identified for the first unit in-

---

[9]Raytheon E-Systems briefing, "HAE UAV CGS Technical Baseline and Program Plan," August 1996.

[10]The MCE schedule driver was the delivery of the communications subsystem from L-3. MCE delivery was to be 60 days after communications delivery to E-Systems. The original schedule had that subsystem arriving at E-Systems in October 1996; actual delivery was July 1997.

cludes more tasks than the recurring cost of the CGS itself. The MCE's cost is driven both by the supercomputer-type processing that is required for the data and information coming from the UAV into a format that theater assets can use by the ground-communications gear.

Figure 5.1 shows the evolution of CGS costs. As discussed above, the program was baselined March 1997 in Amendment 11, which definitized the initial agreement. Up to that point, work had proceeded on CGS at a relatively low level under a not-to-exceed (NTE) amount.[11] The original estimate for CGS (CLIN 0001 in the amendment) was $25 million; as of July 98, the cost had grown to $34 million, a 32-percent increase. The changes required to accommodate the more recent configurations of the air vehicles created this growth. The additional growth is because of activities added to the original statement of work, including studies and analyses, long-lead procurement for a second CGS, and Phase III planning activities. The total value of the CGS Agreement as of July 1998 was $57.9 million.

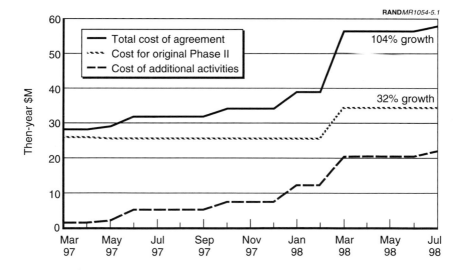

Figure 5.1—CGS Cost Track

---

[11]Work on the Global Hawk ground segment, which forms the basis for CGS, was funded through the TRA Agreement.

The FY98 budget will require that E-Systems take a minor funding reduction. E-Systems sees this simply as a cash-flow inconvenience, because the government eventually will have to provide enough funding for two LREs and two MCEs if the flight-test programs are to be conducted with the air vehicles planned.

SUMMARY OF ISSUES AND OUTCOMES

We have been unable to define exactly what drove the CGS requirement. The rationale appears faint. The two air vehicles have very different technical characteristics, including payloads and software, because of their independent development. Their operational capabilities, mission profiles, and CONOPS are also different, suggesting that they may operate from different bases; the DarkStar has less range and endurance.

One incentive for the CGS was the broader DoD context of reducing duplication in systems as a way to save money in the decreasing budget environment. "Commonality," a buzzword at the time, was encouraged at DoD's highest levels, particularly by the Vice Chairman of the Joint Chiefs of Staff. Thus, to some extent, the HAE UAV's adoption of a CGS was more political than technical or operational, offering increased DoD and congressional support as a result.

As of July 1998, the ground-segment portion of the HAE UAV program appeared to be on track. No serious cost increases or schedule delays had occurred. The 32 percent cost growth in the CGS 1 "upgrade kit" apparently came from changes in air-vehicle configuration, not problems in the CGS itself. The additional growth in the value of the CGS Agreement derives from additional scope, particularly the addition of the second CGS. From a technical standpoint, the CGS would clearly not be fully integrated, at least in the initial versions, but instead would provide co-location for the air-vehicle ground segments and limited commonality in hardware and software. A few potential performance problems began to surface in mid-1998—the LRE's ability to demonstrate simultaneous command and control of both air vehicles and the MCE's ability to demonstrate imagery dissemination to exploitation systems. These problems appeared to be on the way to satisfactory resolution.

Chapter Six

PERSPECTIVES ON THE ACQUISITION STRATEGY

This chapter presents the perspectives of both the contractors and HAE UAV JPO personnel on the acquisition strategy employed in the program. We focus on the innovative attributes of that strategy—ACTD designation, IPPD process, Section 845 OTA, and UFP—while discussing other elements as well. The information presented here is based on a series of contractor and JPO discussions held during Phase II.

While the contractors generally agreed on the advantages and disadvantages of the acquisition strategy and its effect on the program, the JPO tended toward a different perspective. Interestingly, the contractors and the JPO agreed completely regarding the "facts" and important events during Phase II.

The primary discussions with the three prime contractors—TRA, LMSW, and E-Systems—took place in October 1997, with some follow-up in subsequent months. Discussions with the JPO were continuous throughout Phase II.

PROGRAM MANAGEMENT AND STRUCTURE

Contractors were generally positive about the program structure and acquisition strategy. The phase lengths were seen as adequate, despite the accelerated schedule, and the transition from one phase to the next preserved continuity of effort. The contractors liked the management and technical freedom, viewing it as an improved environment for conducting business with the government. For the most part, the JPO was comfortable letting the contractors take the technical lead. The increased contractor design and management responsibility posed a significant challenge to the JPO, however, particularly regarding building the consensus necessary to convince the contractor to do things differently.

The contractors indicated a perceived competition for funding among the program components, but did not emphasize this as a problem. In fact, the JPO wanted the air-vehicle contractors to feel this competition, because both segments ended up as single-source contracting relationships.

Emphasis on the use of COTS was important; the contractors believe that COTS is a useful tool for reducing costs and cycle time, but integration remains a difficult, and often underestimated, task. Additionally, reliability and supportability concerns exist regarding COTS associated with both the military operating environment and the commercial orientation of the suppliers. These concerns have not yet been adequately addressed, although the government and the contractors recognize the potential problems.

We found two COTS-related issues. First is the "vanishing vendor" syndrome (e.g., Texas Instruments' C-80 processor chip). What happens to the design configuration when the original vendor no longer produces or supports a component? Commercial manufacturers are constantly changing their designs, and they are not required to notify customers when a particular item will be discontinued. Under a traditional military-specification (MILSPEC) procurement, the vendor is required to notify the government of the pending termination of a part's manufacture, and the government can then make a "lifetime buy" of the part, procuring enough spares to ensure supportability for the expected lifespan of the weapon systems using it.

Second, COTS applications are difficult because the operational environment is non-COTS. Even with the pressurized payload bay in the Global Hawk, the environment for its components is not necessarily "commercial." The 27,000-foot operating altitude, plus temperature changes and vibrations inherent to flight, are environmental conditions that commercial parts are infrequently designed to withstand. The part vendor has little incentive to build a new military design because the market is small. The small buy quantities give contractors no leverage with suppliers; thus, replacements become unavailable as commercial designs evolve. Even integration is difficult. However, COTS technology is cheaper up front and reduces the development cycle.

Two of the three contractors established separate business units for the program. In both cases, they obtained reduced costs and higher productivity and quality; they also cited improved accountability and the enhanced commitment of employees as benefits of this approach.

The transferral of management responsibility to the Air Force is expected to be smooth because Air Force personnel from the WPAFB projects office have been working the program for years. However, the contractors anticipate changes in both management style and perhaps in program focus; in particular, the Air Force may relax the UFP constraint and want payload changes.[1]

ACTD DESIGNATION

The concepts and definition of an ACTD may have been the source of some confusion early in the program. At least one contractor presumed that an ACTD was similar to an Advanced Technology Demonstrator (ATD), with a focus on demonstrating technical feasibility, and planned its program accordingly. In this case, the original agreement clearly states that the contractor will produce two "proof of concept" flight vehicles, a term that historically has been associated with technology demonstration programs. ACTDs comprise more than this.

Additionally, some contractors perceive that the ACTD concept has evolved, complicating program execution because goals change. We can find only limited support for this view. The earliest program documentation (draft solicitations, management plans) defines the program and ACTD concept in detail and has remained consistent. ACTD guidelines have changed somewhat over time, the most relevant being the use of the term "mature technologies" in more recent guidelines in place of "maturing advanced technologies." While the

---

[1]Indications of this are already apparent. See Fulghum, David, "USAF Takes Over Unmanned Programs," *Aviation Week & Space Technology*, 5 October 1998, p. 106.

difference appears small, it raises the issue of acceptable risk. The notion of mature technology is perceived as translating to a lower-risk program, with risk shifted to integration of subsystems rather than development of those subsystems. We also observe that ACTDs have become more success-oriented than originally intended. Originally, cancellation of an ACTD was not intended to represent a program failure, but rather one of several possible outcomes; participants still learned technologies and operational concepts.

Little effort is spent on reliability and maintainability of ACTDs. Some contractors suggest that substantial investment will be needed in these areas, perhaps more than if they had been considered in the early design stages.

SECTION 845 OTA

Contractors approve of the increased flexibility to execute the program that Section 845 OTA provides. Specific benefits include reductions in auditing and reporting, easier access to preferred suppliers, elimination of nonessential overhead functions, and the use of internal management processes. Cost savings are probably associated with these benefits but their calculation is not clear.

One contractor estimated the cost savings resulting from Section 845 OTA. They included elimination of paperwork, simplified cost proposal, limits on deliverable data, simplified and streamlined contract management, limited program performance reviews, and no MILSPECs. The direct and indirect savings associated with these categories came to 23 percent of the total cost of the program.

In another example, a contractor used the DAPCA 4 (RAND model) and Price H models to estimate costs under a traditional procurement process. The modeling exercise said that Phase II should cost between $400 and $600 million, given program and system characteristics, and take four to six years. Section 845 cut cost and schedule almost in half.

Contractors pointed out that executing the program under Section 845 OTA does not necessarily result in a relaxation of standard company policy and practice. Some effort is usually required to obtain waivers for internal procedures.

IPPD

The application of the IPPD process and IPTs varied among the contractors. The Global Hawk had a formal IPT structure, outlined initially in the Phase II Solicitation, but DarkStar has never been structured that way. CGS also has an IPT structure, originating from the Tier II+ Ground Segment IPT. Nevertheless, all the contractors comfortably operated as product teams. The JPO maintained the IPT structure of the Tier II+ throughout Phase II, but this structure did not always precisely parallel the contractor organization.

Both the contractor and the JPO pointed out that the IPTs were not really "integrated": they did not include at least one member from all participating organizations and of all relevant functions. The cost of executing IPTs to conform with the pure theoretical view would be very high, and none of the participants felt that the lack of integration was a problem.

The combination of Section 845 OTA (implying contractor design responsibility) and the IPPD process allowed a close and positive government-industry relationship, characterized by open communication and cooperation, to evolve. The IPT structure was fundamental to this relationship. It required JPO personnel to participate substantively, rather than just reviewing formal documentation, and allowed them access to the same information that the contractor saw. Thus, the JPO had good insight into program status and problems, albeit not via formal documentation. Both the contractor and JPO point out that this close working relationship, and the need for the government to convince the contractor of the need for change rather than issuing a directive, represents a fundamental cultural change, which both initially took some time to assimilate.

Because there were only three prime contractors with whom to interact, the small program office did not become overextended in Phase II. The contractors perceived that the small JPO was able to manage the program effectively and that the technical quality of JPO personnel was high. The JPO also contained the right skill mix for the Phase II activities.

Despite the overall positive view, the IPT structure and IPPD process did not always work as intended. In theory, members of an IPT are equal, and must reach a consensus in order to make a decision. In practice, all members are not equal. In some cases group dynamics and long-standing cultural norms allowed the government to dominate decisionmaking. A strong government personality and the historical defense contractor culture of deferring to the government are important factors here. This problem raises the issue of government accountability for decisions, particularly additional costs. Thus, the personalities and attitudes of both government and contractor personnel can affect how well an IPT functions.

A potential conflict exists between Section 845 OTA and contractor authority over design and management on the one hand, and government participation on IPTs on the other. In the IPPD process, all team members are equally responsible for the outcomes of decisions. The means by which government accountability is maintained under this acquisition strategy is not clear. This is a serious concern that requires investigation.

Contractors also perceived that government IPT members often operated in an oversight role, rather than as a technical participant. Government technical participation was low in some areas. A proper balance must be struck between the technical and oversight roles of government.

A final critique of IPTs is that implementation can be carried to an absurd extreme—IPTs with only one or two members. When there are many small IPTs, significant communication and coordination problems may challenge program management.

A potential conflict of interest exists when the government representative is a SETA contractor. First, because SETA organizations get paid for studies and analysis, their incentive may be to call for additional analysis rather than make an action-oriented decision. Second, because some SETA contractors are also prime contractors in other programs, and may compete on future programs, inappropriate transfer of proprietary information may occur. Last, one contractor declined to allow the in-house SETA person, who would not agree to certain rules of engagement regarding parallel and concurrent reporting, to participate in critical meetings. The result was a lack of insight into that aspect of one program. While these issues were raised in the context of HAE UAV, they are unrelated to the program acquisition strategy and apply more broadly to program management approaches that rely extensively on SETA support.

UFP COST AND PERFORMANCE TRADES

Cost-related issues, including the UFP requirement and the adequacy of nonrecurring engineering funding, were important issues in our discussions.

The contractors and the JPO believe that the UFP requirement is a reasonable way to control costs and to encourage cost–performance tradeoffs. Program participants believe that such a constraint drives creative solutions. However, the contractor must have complete discretion in defining the system—a responsibility that contractors have been pleased to have. However, the amount of analysis that went into defining the $10-million UFP and the NRE funds available to achieve that goal is unclear.

The contractors and the JPO acknowledge that both air vehicles will exceed the UFP by an estimated $1 to $3 million per air vehicle. While this information has been widely briefed among participating organizations, the government states publicly that the UFP remains achievable. And, in fact, the contractors have identified cost reductions that would enable the UFP to be achieved in the future.

Production costs will exceed the UFP for a number of reasons. Assumptions underlying the UFP have been violated, including production gaps and transitions to subsequent phases, subsystem initial costs, and cost improvement. Also, the contractors assert that development has been underfunded. Additional NRE investments in production engineering and tooling need to be made in order to achieve the $10-million UFP.

Perhaps more important, insufficient cost-performance tradeoffs are being made to meet the UFP target. Some compromises in performance have been made, but no functionality has been dropped. These tradeoffs are being neglected largely because of the perceived ambiguity in the definition of military utility. When management transferred from DARPA to the Air Force, military utility had not yet been defined such that the contractors could translate it into system capabilities. Hence, the contractors have avoided trading off functions for fear of inadvertently dropping capabilities the users wanted most. Some contractors also felt that, despite the JPO's intent and the wording of the solicitation, they were not free to make performance trade-offs.

The government imposed some subsystems resembling the "requirements creep" noted in traditional acquisition programs. The government was not adding superfluous systems; rather, they enabled or enhanced operational capabilities that DARPA did not understand well.[2]

An additional cost-related issue relates to the contractors' contention that they are accomplishing more with less money than previous programs (e.g., Have Blue, Tacit Blue). They assert that the systems being developed will be much more missionized than the technology demonstrations of past programs and the cost to the government is less. If this is true, the government's ability to glean more from its investment may be because of a declining federal R&D budget and fewer new projects, rather than a result of the acquisition strategy.

Another cost and performance issue involves the lack of a formal change procedure. The contractors pointed out that while not having a formal process makes it easier to incorporate design changes, the party responsible for the changes' consequences can be hard to determine. This issue goes back to government decisionmaking accountability in the IPT structure.

OTHER ISSUES

The contractors and the JPO agree that the event that affected the program the most was the crash of DarkStar AV-1 during takeoff on 22 April 1996. The following is a list of contributing factors.

- The contractor, LMSW, pushed the schedule beyond the already accelerated plan. As discussed in Chapter Four, the contractor was pushing in order to demonstrate its superior ability in advanced technology demonstration programs—not because of a

_____

[2]For example, adding position lights and voice radio capability to DarkStar.

contractual obligation. The contract included a minor cost incentive to accelerate the schedule.

- The contractor did not build a sufficient database on the aerodynamics of this configuration, nor did it perform sufficient wind-tunnel tests beforehand. The relevance of prior experience was significantly overestimated.

- The contractor program management exercised poor judgment regarding risks, particularly when its own engineers reported unexplained discrepancies between modeling and simulation data and the data from the first flight.

The crash brought an increase in risk aversion throughout the entire HAE UAV program. It precipitated increased reviews and more conservative design decisions, which affected the cost and schedule of both the Global Hawk and CGS, but whether it was a dominant factor is uncertain.

While normal oversight and review procedures were not in effect for DarkStar, the program office knew of the poor aerodynamic database and flight-data discrepancies prior to the go-ahead for the second flight. The JPO wanted to delay the second flight, but was overruled by senior government managers. The current contractor management asserts that a four-month extension of development and testing prior to the first flight would have generated the data to identify and resolve the technical problems.

Another issue is the air-vehicle contractors' belief that the CGS effort is premature. It increases program risks and forces the government to act as system integrator. It also is inappropriate for a ground-system contractor to affect the design of air-vehicle mission control. Operational issues question the value of the CGS: the Global Hawk and DarkStar have completely different capabilities (range and endurance) and mission profiles (low- versus high-threat environment), and seem likely to be operated independently. The JPO does not find the effort premature, but agrees that risks are increased and admits its role as system integrator. The office points out that the decision to develop a common ground segment early in the program was made in part to gain DoD and congressional support by reducing apparent duplication in the program. When the program began, commonality was a major theme in system development planning.

From a different perspective, the difficulty in developing an integrated common ground station is increased because the two air vehicles, and associated flight and mission control systems, were designed independently. They have no commonality in hardware or software. The initial CGS effort is like a common packaging effort, with the system integration applied when practical, but still somewhat limited.

Chapter Seven

INTER-PROGRAM COMPARISONS

This chapter compares the HAE UAV program's cost and schedule with other programs. The comparison is problematic in that the HAE UAV program is unique; inter-program comparisons will be misleading unless one controls for that individuality in the analysis. More important, the singular characteristics of acquisition programs generally prevent definitive comparative statements regarding program outcomes and efficiency of the process. Finally, this comparison must be regarded as a work in progress, because the HAE UAV ACTD has yet to be completed.

PURPOSE AND LIMITATIONS OF INTER-PROGRAM COMPARISONS

We compare the HAE UAV program with other programs to gain better understanding of the potential costs and benefits of its innovative acquisition strategy. Conceptually, we would prefer to compare the cost and schedule of the HAE UAV program, as executed, with a "nominal" HAE UAV program conducted using traditional acquisition policies and processes. Available data do not support synthesis of a nominal program. As a result, we must assess differences qualitatively; we cannot measure them quantitatively.

We have gathered cost and schedule data from select programs. We include aircraft, cruise missiles, and other UAVs that experienced Milestone I, or its equivalent, in 1970 or later. This time frame enables relative consistency in acquisition policy.[1] We use actual data where possible, but include estimates where necessary and identify them. Only new development programs for which data are available made the comparative program set.[2]

Each acquisition program is unique; that uniqueness must be considered in the analysis and interpretation of results. Our experience analyzing acquisition programs suggests that critical factors include the technical challenge and complexity of the system, the program structure, the external environment in which the program is executed, the players involved, and the activity content of each phase.

Activity content is critical. We define it as the specific developmental activities that are performed in a given program phase. Activities might include how deeply supportability concerns penetrate the design phase; logistics planning; the type and amount of testing conducted at the system and subsystem levels; the thoroughness, type, and number of program reviews; and the roles of review participants. Activity content of each program phase ultimately determines the relative maturity of the system at phase end. Given the singular activity content in each phase of each program, any comparison of programs will inherently include systems at different levels of technical and operational maturity.

We believe that the activity content of the HAE UAV ACTD program is far less comprehensive than a traditional EMD phase, but more comprehensive than a typical technology demonstration program. The phases of the HAE UAV ACTD program have counterparts in the traditional process. Phase I of the Tier II+ program is roughly equivalent to a concept-exploration phase (Milestone 0 to Milestone I). Phase II for both the Global Hawk and DarkStar includes activity content that is more comprehensive than a typical demonstration or validation phase (Milestone I to Milestone II). The user evaluation in

---

[1]The current acquisition-policy regime dates to December 1969 and the issuance of the first DoD Directive 5000.1. Though the specifics of the acquisition process have often changed since then, the policy has remained consistent.

[2]We would like to include more programs over a larger set of metrics, but data are not readily available. Some metrics that would be useful, but are not included here because of data unavailability, include program-office staff sizes, number of reviews, number of audits, CDRL counts, and government resources expended. We are forced to use simple cost and schedule metrics.

Phase III of the HAE UAV program is roughly equivalent to the initial operational test phase in a traditional program. Aircraft fabricated during Phase III of the HAE UAV program are expected to be the rough equals of developmental aircraft fabricated at the end of an EMD program (i.e., full-system configuration on limited production tooling).[3] The HAE UAV ACTD program combines and sequences activities in a significantly different way than a traditional acquisition program does. The implication is that, at any particular "milestone," the Global Hawk and DarkStar have different levels of maturity from other programs at equivalent "milestones."

COST

Figure 7.1 shows the expected Global Hawk and DarkStar ACTD costs and compares them with the costs for the Compass Cope (YQM-98A and YQM-94A) program and the Light Weight Fighter (LWF: YF-16, YF-17) technology demonstration. Compass Cope data include the actual cost of the prototyping phase and the proposed development program. LWF program costs are for the single phase in which the prototype aircraft were designed, fabricated, and tested. Costs for the third and subsequent air vehicle in the Compass Cope and HAE UAV programs are estimated based on available program documentation.

The costs for Phase II of the HAE UAV program are roughly equivalent to the costs of the LWF program, and considerably higher than the prototype demonstration in the Compass Cope program. The aggregate activity content of these early phases for all three programs is comparable. Each program included the design, fabrication, and test of two air vehicles of each type (e.g., two YF-16s, two Global Hawks, etc.). In the HAE UAV program, however, complete system prototypes were constructed. In both comparative programs, the flying hardware included only airframes and engines. In the Compass Cope program, payloads were excluded. The total cost of

---

[3]The activity content of programs changes over time as more is learned about the system and thus attention can be focused on critical areas. The difference in the HAE UAV program is that many activities have been removed to maintain the overall budget and schedule. In a traditional acquisition program, activity content would not be compromised to the extent observed in the HAE UAV program, and cost increases and schedule slips would more likely occur.

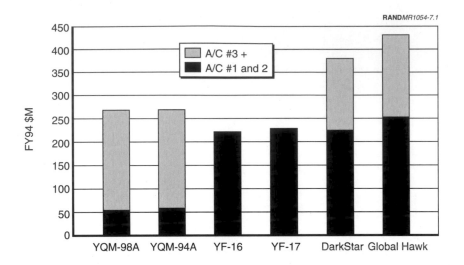

Figure 7.1—HAE UAV Demonstration Cost Comparison

either Global Hawk or DarkStar is somewhat greater than the estimated total development cost of either the Boeing (YQM-94A) or TRA
(YQM-98A) versions of Compass Cope. In the LWF program, avionics
and weapons management were excluded. Only the HAE UAV
program emphasized demonstrating end-to-end "connectivity."

Figure 7.2 provides a different perspective. The same costs for the
HAE UAV ACTD program and TRA's Compass Cope UAV are shown.
The full development costs of the F-117, F-16, and F/A-18 are also
shown. The demonstration/validation portion of the F-16 and
F/A-18 is the LWF program. The Have Blue technology demonstration is counted as a demonstration/validation equivalent for the
F-117. All of these programs used streamlined acquisition procedures, at least in their early phases. The top of the bars in Figure
7.2 represents the costs through delivery of an operationally useful
system. For the F-117, about half of the FSD costs shown were
incurred after the first operational delivery and represent
supportability and other upgrades.

The maturity level of each system in Figure 7.2 is unique. The
manned aircraft data include the full production and support engi-

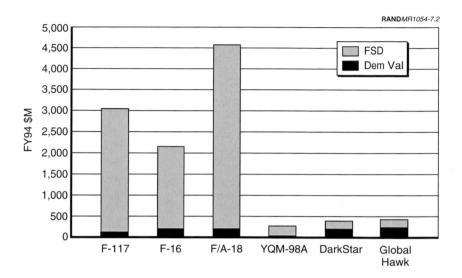

Figure 7.2—HAE UAV Development Cost Comparison

neering costs for a suitable operational system. This required evolution of the design such that it met the desired performance specifications. The contractors also developed a manufacturing capability for efficient serial production. The HAE UAV program does not include these activities. At the end of the HAE UAV ACTD program, the Global Hawk and DarkStar will likely require an EMD-type phase if program participants seek to significantly improve reliability, maintainability, producibility, and overall system performance. Despite this expectation, the HAE UAV ACTD cost to design, develop, and test a potentially suitable configuration is small in comparison to the fully developed programs.

If the ACTD process identifies needed engineering improvements, then a follow-on EMD program may include a subset of the activities found in a typical EMD program. The cost of such a "tailored" development phase would therefore be a fraction of a typical EMD phase.[4]

---

[4]This effect cannot be determined until the ACTD program is completed and a decision is made regarding future development and production.

A more detailed comparison of the Have Blue/F-117 and the HAE UAV programs provides additional insight into relative costs. Both programs used streamlined management with few hard specifications and exceptional delegation of authority to the contractor. Both were exempt from most of the usual oversight and reporting requirements. Both started with an emphasis on technical performance that evolved to an operational emphasis, though only the HAE UAV was planned that way. The cost of the Have Blue technology demonstration program was approximately $170 million (FY94 dollars) and included 88 flight tests over 17 months. In comparison, the Global Hawk Phase II effort cost approximately $250 million (FY94–FY98 dollars) and is planned to include 16 flights for a total of 191 hours. The Global Hawk appears to cost slightly more, but it will accomplish more in design for operational suitability. While the HAE UAV flight-test program includes fewer flights than the Have Blue, the flight tests are more comprehensive because of the demonstration of an integrated mission system.

The effects of the streamlined management processes used in the F-117 program on outcomes were mixed. They had no apparent effect on overall cost or schedule; the F-117 is comparable to most contemporary fighter-aircraft programs. However, the government required fewer resources to manage the program. Initial operational suitability was poor, a result directly traceable to the development strategy (supportability concerns were not emphasized).[5]

SCHEDULE

Measurement and comparison of program schedules is a deceptively complex task that must address several ambiguous definitions. The fundamental problem is that, regardless of the terminology used, the identification of precise dates is elusive for key milestones such as program start, start of engineering development, and first operational delivery. In addition, as discussed above, the level of a program's maturity at any particular milestone will be different than that of any other program because of the activity content and execution

---

[5]For a full analysis of the F-117 program, see Smith, Giles K., Hyman L. Shulman, and Robert S. Leonard, *Application of F-117 Acquisition Strategy to Other Programs in the New Acquisition Environment*, Santa Monica, Calif.: RAND MR-749-AF, 1996.

of the preceding phases. Thus, one should interpret the results of a comparative schedule analysis, like those of the cost comparison above, with caution.

We use the following definitions in our analysis:

- *Program start*: This is usually a pre–Milestone I activity. It is the earliest point at which a rudimentary operational concept exists, broad performance goals have been established, and a decision has been made to build and test one or more systems.

- *Milestone I*: The earliest formal program initiation. May be a formal Milestone I approval, establishment of a program office, start of demonstration or validation activities, or a contract award for such activities.

- *Milestone II*: Beginning of engineering development. May be the formal Milestone II approval or contract award for EMD.

- *First operational delivery*: Delivery of the first production system, in final configuration, to the operational forces for operational use.

Given a choice, we use contract-award dates rather than formal Milestone approval dates, because they better reflect the beginning of specific activities in a phase. By definition, the first operational delivery in an ACTD occurs after the ACTD program is complete.[6] The definitions above allow us to create three distinct phases:

- Phase 0: time from program start to Milestone I

- Phase 1: time from Milestone I to Milestone II

- Phase 2: time from Milestone II to first operational delivery.

All phases are measured in months. Phase 1 corresponds with the DODD-5000.1 definitions, while the other two phases do not corre-

---

[6]For both the Global Hawk and DarkStar, this was estimated to be 18 months after the force-mix and military-utility decisions.

spond with any of the traditional acquisition measures used by DoD.[7] The sum of the three phases equals total program length.

Figure 7.3 shows the length of these phases for a small number of selected programs.[8] The key observation is the variation in the length of any phase across these programs. The data suggest that no typical length exists for any given phase, even for this relatively homogenous group of programs. The DarkStar and Global Hawk schedules are among the shortest in this group.

We divided the group of programs into three categories: ACTDs, programs that produced a prototype prior to Milestone II, and programs in which pre–Milestone II activities consisted mainly of studies and analyses. Table 7.1 shows the average length of each phase for each category.[9] This sample size is too small to make inferences regarding differences, but if this pattern were to hold in a more robust sample of programs, it would suggest that a shortened period of analysis (Phase 0) and system-level hardware demonstrations prior to engineering development could shorten the total time required to

Table 7.1

Average Phase Length for Selected Programs

	Average length (months)			
Category	Phase 0	Phase 1	Phase 2	Total
ACTD	13	24	34	66
Prototype	17	38	63	109
Studies	29	26	126	121

[7]These measures are consistent with past RAND work. See Drezner and Smith, *An Analysis of Weapon System Acquisition Schedules*, Santa Monica, Calif.: RAND, R-3937-ACQ, December 1990; Smith and Friedmann, *An Analysis of Weapon System Acquisition Intervals, Past and Present*, Santa Monica, Calif.: RAND, R-2605-DR&E/AF, November 1980; Drezner et al., *An Analysis of Weapon System Cost Growth*, Santa Monica, Calif.: RAND, MR-291-AF,1993; and Rich and Dews, *Improving the Military Acquisition Process*, Santa Monica, Calif.: RAND, R-3373, February 1986.

[8]For almost half of these programs, the length of Phase 2, and thus the date of first delivery, is estimated. This introduces another source of potential error.

[9]As discussed previously, the criteria for inclusion in the comparative set of programs, and the relatively small number of new programs begun in recent decades, severely constrain the sample. The sample size is too small for both the ACTD (n=5) and studies (n=3) groups, making conclusions unreliable.

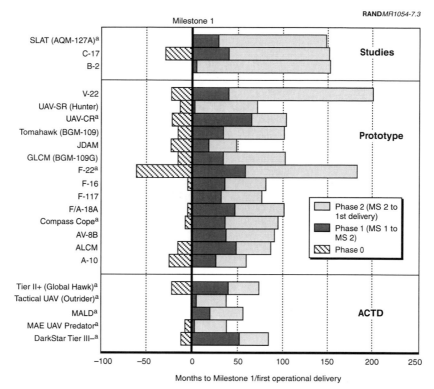

Figure 7.3—Schedule Comparison Overview

field a new weapon system. As discussed above regarding cost, Global Hawk and DarkStar may be examples of this schedule benefit, depending on the activity content of any further engineering development work.

Figure 7.4 shows additional detail on selected programs.[10] We have added data on the time to first flight of the prototype and time to first flight of an EMD test aircraft. The length of the phases for Global

_____

[10]The length of Phase 2 is estimated in all but two cases, again leading to potential error.

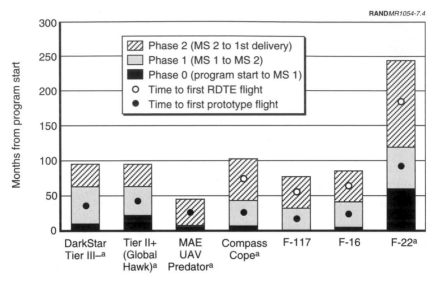

aFirst operational delivery estimated.

Figure 7.4—Detailed Schedule Comparison of Selected Programs

Hawk and DarkStar are roughly comparable to those of the F-117 and F-16, as well as to the estimated schedule for Compass Cope. Interestingly, the time to first flight of the prototype for Global Hawk took somewhat longer than for either the F-16 or F-117, while DarkStar's timing is more comparable.[11] The F-22 schedule is extremely long compared to any of the other programs shown.[12] However, Phase 2 in the F-22 program (the ATF prototype demonstration) is roughly the same length as Phase 2 in Global Hawk and DarkStar.

The activity content of each phase drives schedule length. Programs at nominally the same "milestone" can therefore be at different levels

---

[11]Of course, DarkStar crashed one month later largely because of the lack of understanding of the aerodynamics of the design. See Chapter Four.

[12]The selected program start date was November 1981, corresponding to DoD approval of an ATF statement of need and the establishment of a SPO in ASD. The subsequent period of studies and analysis lasted several years. Milestone I was October 1986.

of maturity. The Global Hawk and DarkStar schedules resemble those of other streamlined programs, and are appropriate for comparing their activity content to that of traditional programs.

CONCLUSIONS

The brief inter-program comparison presented in this chapter suggests that, for the most part, the cost and schedule of the HAE UAV program are reasonable, given the program's activity content. We can make no valid direct comparisons, but the ACTD program appears more robust than a typical demonstration/validation program, and less robust than a minimal[13] development phase. However, compared to past technology-demonstration programs, the HAE UAV ACTD appears to achieve more for about the same cost, though final assessment awaits the completion of the ACTD.

Schedule metrics—program length, phase length, and time to first flight—are nominally similar to other programs. Development costs are consistent with a typical demonstration/validation phase, but will result in a more mature system. Additionally, users can participate earlier than in a traditional operational test phase.

These comparisons, however, are not how the HAE UAV program should be assessed. Rather, the value of the HAE UAV program relative to other programs (and therefore the value of its acquisition strategy compared to a traditional strategy) should be assessed in two ways:

• whether the program expended time and money in only those areas that were needed to demonstrate military utility. In other words, the acquisition strategy should be judged by whether it facilitated removal of expensive and time-consuming activities that are normally included in a traditional development program. Some of these activities may have improved quality, but they might not have been cost-effective in the current acquisition environment.

---

[13]"Minimal" refers to the activity content of a program that results in an operationally suitable system.

- whether the knowledge and experience gained accelerates the schedule and reduces the costs of any future engineering development. Hopefully, future work can be more focused on specific needs identified by the user.

Neither of these propositions can be answered until completion of the ACTD program.

Chapter Eight

OBSERVATIONS ON PHASE II

We cannot offer a definitive analysis of the effect of the acquisition strategy on outcomes this early in the HAE UAV program. We can, however, offer a preliminary assessment. This chapter presents that assessment and draws lessons from the HAE UAV program's experience so far that can be applied to the broader acquisition community.

LINKING ACQUISITION STRATEGY TO OUTCOMES

The effect of the innovative attributes of the acquisition strategy—ACTD designation, Section 845 OTA, IPPD process, and UFP/Performance Goals—is complex, prohibiting simple explanation. Each element of the strategy has advantages (benefits) and disadvantages (costs). Complex interactions take place among the various elements, making it difficult to distinguish the effect of one attribute over another. These interactions produce both positive and negative effects. Other elements of the acquisition strategy unrelated to the innovative attributes also affect program outcomes.[1] Lastly, the acquisition environment itself is too complex for any strategy to resolve.[2]

---

[1]For example, funding profiles, adequacy of funding, and source-selection decisions.

[2]Previous RAND research, both published and unpublished, suggests that acquisition strategy is only one set of factors affecting program execution and outcomes. External political influences, perceived urgency of the threat, budget environment, funding stability, and turnover in key program and home-agency personnel all present challenges. See Drezner, Jeffrey A., and Giles K. Smith, *An Analysis of Weapon Acquisition Schedules*, Santa Monica, Calif.: RAND, R-3937-ACQ, December 1990;

While traditional program outcomes—cost, schedule, and performance—are important measures of program success, the HAE UAV acquisition strategy employed was meant to explore nontraditional outcomes as well. These include the quality of the government-industry relationship; communication flow, flexibility, and responsiveness; reduced non-value-added burden; faster deployment of a minimum capability; and user involvement. The relative success of the strategy should include a measurement regarding these outcomes as well as the more traditional ones.

No silver-bullet package of reforms exists that guarantees a "good" outcome. The strategy used in the HAE UAV program facilitated many outcomes believed to be desirable by the government acquisition community and industry. However, it also contributed to less desirable outcomes. While future applications of the strategy can adjust execution according to lessons learned, we find no inherent flaw in the strategy itself.

We discuss the observed advantages and disadvantages of the HAE UAV program's innovative acquisition strategy under the four key initiatives that comprise that strategy: ACTD designation, Section 845 OTA, IPT structure, and UFP and performance goals.

ACTD Designation

An ACTD designation has two key effects on a program. First, it bounds total program length; demonstrations must occur in three to five years. Second, it provides for user involvement early in the program, including during the design and development phases.

We believe that the bounded program length resulting from the ACTD designation is potentially insufficient to determine military utility for both the Global Hawk and the DarkStar. As noted earlier, despite significant changes in interim milestones, the ACTD program completion (and end of funding) has remained fixed. Development and test activities have been significantly reduced to meet this schedule. In particular, the user demonstration phase has been re-

Rich, Michael, and Edmund Dews, *Improving the Military Acquisition Process: Lessons from Rand Research*, Santa Monica, Calif.: RAND, R-3373/AF/RC, February 1986.

duced by half, limiting the scope of user activities and range of conditions under which the systems can be evaluated.

Accelerating the early phases promotes increased risk in later phases. The ACTD designation led to an accelerated schedule, especially in the early phases, and probably increased the risk in later phases. The accelerated pace is responsible for some of the problems experienced.

A more fundamental problem for the HAE UAV ACTD was the lack of an agreed-upon definition of military utility. In a sense, users were expected to "know it when they saw it." This complacency was unjustified; it failed to recognize the ensuing negative effect on the contractors' and program office's willingness to make performance-cost trades.

Incorporating the potential for early user involvement in the design, development, and test phases means rearranging the traditional sequence of development activities. In particular, less effort is expended on support and producibility efforts until the user has tested and evaluated a final design. This can be more easily accomplished in an ACTD program, and to the extent that opportunities are exploited, may be a significant benefit of this program structure. Early user participation may result in decreased costs and accelerated schedules in future development activities. In the HAE UAV program, this opportunity was missed: users were informed of program status and direction, but provided little input.

The ACTD designation streamlined the decision and oversight process, and, in combination with the mandated short schedule and early user involvement, resulted in increased design flexibility and the potential to demonstrate a new capability faster than in a traditional acquisition program. The ACTD designation enabled accelerated development and testing of full-system prototypes. The Global Hawk took 34 months to first flight and the DarkStar took 22 months—a relatively fast pace for complex aircraft systems.[3]

---

[3]The planned times to first flight were short by any standard: 20 months for Global Hawk and 15 months for DarkStar. Given the DarkStar crash beginning its second flight, and the subsequent delays in both programs, the original schedules were too short.

Section 845 OTA

Section 845 also produced a mixed effect on the HAE UAV program. The increased contractor design responsibility and management authority resulting from the way Section 845 OTA was implemented apparently decreased decisionmaking time and increased flexibility. The contractors owned the process to a greater degree than in traditional processes and could move faster. Reporting and other management processes were internal to the company, rather than imposed by the government. A disadvantage of this structure is the lack of a formal contract-change process; though disputes are less common, they may be more difficult to resolve.

The use of contractor-based management processes (reporting, auditing, accounting, etc.) reduced overhead costs. Levels of review were eliminated. One contractor estimated that overhead costs were reduced up to 50 percent from what they would have been without Section 845 OTA.[4] Other contractors provided qualitative estimates of how much overhead costs were reduced because of the lack of government auditing, the elimination of complex reporting and accounting processes, and the ability to pass these changes through to lower-tier suppliers.

Overall, we believe that the relatively low costs for the basic design and development efforts for the two air-vehicle systems (about $200 million each for two air vehicles and a short test program) can be substantially attributed to management of the program under Section 845 OTA. In comparison to past experience (e.g., Have Blue/F-117, U-2), this is inexpensive, given that the intended result is an integrated system suitable for operational demonstration, not just a technical-feasibility test.

In contrast, the flexibility inherent in Section 845 led to a lack of discipline in the program's systems-engineering approach. That lack of discipline is indicated by the concurrent development of the systems and their specifications, software development problems, and inadequate attention to systems integration. One JPO manager stated that while the full application of the government's standard systems-

---

[4]The documentation supporting this estimate is scarce; the validity of the estimate could not be determined.

engineering process was not necessary, tailoring that process to the program would have been an improvement on the processes that were actually used.[5]

IPPD Process and Use of IPTs

While the IPT structure used in the program facilitated timely insight, the government lacked a mechanism to directly affect decisions.

Though strained at first, the government-industry relationship eventually evolved into a positive interaction that the JPO and the contractors both characterized as an improvement. The initial tension was due in part to the need to adjust to the radically different acquisition strategy, as well as at least one personality-driven problem. While traditional oversight processes were reduced, the participation of JPO representatives on contractor IPTs provided timely insight into program status, problems, and solutions, allowing JPO officials to provide timely input into contractor decisions. Information was more candidly exchanged than in a traditional oversight relationship.

There is a potential conflict between the IPPD process, which encourages teamwork, and Section 845 OTA, which encourages contractor responsibility. This conflict manifested at least once in the HAE UAV program during the Global Hawk wing redesign. The issue revolves around the participation of JPO representatives on contractor IPTs when the contractor has decision authority. Under these conditions, who owns the process is ambiguous, especially if the government uses the IPT structure to gather information in support of more traditional oversight and review rather than substantive participation. In an IPT, the team is accountable for decisions; who is accountable when the government is part of the team is unclear. While this conflict may not be inherent in the two processes (IPPD and Section 845 OTA), it is inherent in their implementation. An optimum balance between oversight and technical participation has not yet been achieved.

---

[5]The government systems-engineering process referred to is MIL-STD 1521.

An issue related to the use of the IPPD process is that a small JPO can be overworked. The acquisition strategy placed more of a burden on the small program office than dealing with more firms in the initial competitions. Close interaction with the contractor required extra effort by the JPO officials: participating in a value-added manner is more difficult than reviewing someone else's work. The contractor also felt this burden. Establishing and maintaining a good relationship takes extra effort by all parties.

UFP and Performance Goals

The UFP requirement, with all other system characteristics stated as performance goals, was a more radical approach to system design than the contractors anticipated. Under this approach (combined with other elements of the acquisition strategy), contractors theoretically had complete control over the cost-performance tradespace. As discussed above, inadequate definition of military utility crippled the willingness and ability of the contractor to make cost-performance tradeoffs.

Both the contractor and the government were reluctant to drop functionality as a way of meeting the UFP. The motives behind this reticence were similar: functionality would largely determine utility, so until utility was clearly defined, dropping functionality risked program termination at the end of the user evaluation. We believe that better guidance from the user community would enable the cost-performance tradeoffs expected under this management approach. In particular, user input could be enhanced; more formal participation and improved coordination among potential users could provide improved guidance without constraining the contractor's tradespace. At the least, a priority ranking of desired performance characteristics would have provided valuable information to the contractors during design and development.

In order to achieve UFP, certain NRE activities must be performed and investments made. An impeccable rationale must determine the balance between NRE funding and the UFP for the UFP to be credible and achievable.

To meet an apparent NRE-funding shortfall, the contractors are "creating" additional NRE funding by reducing developmental-activity content (which also minimizes developmental cost growth). The acquisition approach intended for functionality to be traded to meet UFP. The two tradeoffs, allowed by the structure of the agreement, have different implications for system development and technical maturity. The government should define which approach it desires and prohibit the other.

Lastly, while the UFP is likely to be breached in the HAE UAV program, even significant UFP growth should not be considered a failure. A 50-percent growth in UFP for either the Global Hawk or DarkStar (i.e., a $15-million UFP) could still result in a capable and cost-effective mission solution. UFP might also be set as a goal—perhaps the most important goal—along with other system characteristics, thus enabling more effective cost-performance tradeoffs and aligning the expectations of program participants.

General Observations

Problems should be expected in a first-time application of any radical innovation. Would a more traditional approach—increased oversight, mandated processes, larger JPO—have prevented the problems that occurred? Both the JPO and contractors believe that problems would have occurred anyway. On balance, both the JPO and the contractors rate the acquisition strategy highly. In particular, all cited the improved working relationship that evolved between government and industry as beneficial. Cost savings and an accelerated schedule relative to more traditional processes were also cited as key benefits.

We also observe that the high degree of process flexibility inherent in this acquisition strategy requires high-quality JPO personnel, sustained senior-management support within the relevant DoD organizations, and a cooperative contractor willing to accept increased responsibility. Participants must be adaptable and open. All parties must recognize benefits to the approach and acknowledge that a key goal of the program from the outset is to demonstrate that approach and its viability, not just to develop a system.

TRANSITION ISSUES

The HAE UAV program was to transfer to Air Force management at the beginning of FY99. At this transition point, the program still is an ACTD. A second transition, to a more formal acquisition process and operational use, would occur after the military-utility evaluation at the end of Phase III. Accomplishing these transitions is critical to a successful demonstration program.

The original HAE UAV program MoU (dated October 1994) began to lay the groundwork for a successful transition from DARPA to Air Force management. It assigned an Air Force colonel and a Navy captain as deputy program managers. The MoU was routed among the staff of relevant organizations across DoD, and the services and the rationale for the program was based on a validated JROC MNS for RSTA (JROCM-003-90, 5 June 90). The initial MoU also mentioned the principle of event-based timing and associated the transition to Air Force management with the completion of Phase II. The MoU identified the Air Force as the lead agency for Phase III and beyond, and service deputy program managers were intended to transfer with the program, along with other service-specific billets in support. The MoU specified service roles and responsibilities. The structure of the plan—transferring management to the lead agency during the ACTD program—could facilitate successful transition to the acquisition process and operational users.

Transition planning began early. The Aeronautical Systems Command (ASC) JPO at WPAFB was established in November 1995, though logistics issues were being supported as early as August 1995. The ASC JPO was intended as a "shadow" organization to monitor and support HAE UAV progress. In fact, personnel from the ASC JPO have been key in all HAE UAV program activities, and the organization charts from the two JPOs list many of the same personnel. The July 1996 transition plan, signed by DARPA and Air Force principals, outlined the general transition approach, established a group to update the transition plan and resolve issues, and directed the development of a lessons-learned database. Key decisions affecting the program transition require approval of both the DARPA Director and the Commander of the Aeronautical Systems Center at WPAFB. Transition activities were defined in the following areas:

- technical: risk assessment, continuity of technical knowledge
- financial: funding
- contracts: contractual mechanisms
- test and demonstration: residual tasks, assets
- supportability: reliability and maintainability
- programmatic: coordination with related programs.

Recently, senior Air Force managers who will be responsible for the program beginning in FY99 expressed uneasiness with elements of the acquisition strategy used in the HAE UAV program. One cause for this phenomenon is that many of the senior Air Force personnel who were originally involved with the program have moved on. This illustrates the difficulty and importance of maintaining consistency in management approach during and after a transition. Analysis of the transition is an important issue for future research.

LESSONS LEARNED

Section 845 OTA provided tangible benefits to the program office and the contractors, including less-burdensome and more-informal management processes, reduction in overhead costs, and an improved work environment. We believe that Section 845 OTA (Section 804 for the military services) is more widely applicable; entire programs, or portions of larger programs, can be conducted using this authority. The Navy is trying this approach in the DD-21 program. The key to successful use of this increased flexibility will be to strike a balance between increased contractor management authority and more traditional oversight mechanisms.

Future use of the innovative acquisition strategy used by the HAE UAV program office will require an adaptable government management approach to offset contractor weaknesses. While reliance on the contractor for design and management processes can save time and money, the government must intervene when the contractor demonstrates weakness in key areas. A mechanism for this type of intervention should be incorporated into the agreement defining the program and government-industry relationships.

The potential benefits of involving the user both early and in the development process remain untested. One reason for this is that such user involvement is a significant cultural change for the user and acquisition communities, and it bypasses the requirements community almost entirely. We believe that the mechanisms for user participation should be more formalized, accompanied by improved coordination among user groups and a clarification of expectations. Military utility must be defined early in a program to provide guidance to contractors for ranking the priority of potential system capabilities.

Comprehensive planning of the initial program structure, including consideration of program risks, would improve program execution. Risks need to be identified early in the process, and a plan developed to manage them. Most of the technical sources for problems in the Global Hawk and DarkStar programs should have been anticipated early in the program. In our Phase I report, we suggested that UAVs inherently require a lower-risk program structure.[6] The HAE UAV program appears to have demonstrated this, since many of the problems experienced can be attributed in hindsight to the relatively high risk that was accepted, even if it was not recognized at the time.

More fundamentally, the success of an ACTD program depends on whether its selection as an ACTD was appropriate, considering its level of risk. Program offices have a natural incentive to claim a lower level of risk than may be justified; "mature technology" (an ACTD prerequisite) is a hazy concept. We recommend an independent risk assessment prior to designation of a program as an ACTD.

We believe that future programs should evolve beyond the limitation of UFP as the single requirement. Cost, schedule, and performance can all be goals to be traded against each other to achieve an optimal solution to the military mission. Programs can set boundaries for cost, schedule, and performance parameters, but the resulting trade space must be large enough to enable realistic and credible tradeoffs. This flexibility is perhaps appropriate in the near future only for small programs that incorporate new capabilities and concepts, and

---

[6]See Sommer et al., *The Global Hawk Unmanned Aerial Vehicle Acquisition Process: A Summary of Phase I Experience,* Santa Monica, Calif.: RAND, MR-809-DARPA, 1997.

should be tried several times on an experimental basis before being applied more ambitiously. We believe that this approach can better produce a cost-effective capability in a shorter period of time.

EVOLUTION OF HAE UAV PROGRAM AGREEMENTS

The following tables show the evolution of the Agreements for each of the three segments of the HAE UAV program: Global Hawk, DarkStar, and CGS. An Agreement may not be a single document; it often includes a series of Amendments that are added after the initial Agreement is signed. Each Amendment changes only those portions of the original Agreement specifically mentioned; all other articles and provisions remain in effect. The notes associated with each table define terms and describe key changes to the program.

The data included in Tables A.1 through A.3 represent the data available at the time of this writing (August 1998) and portray the evolution of each program segment.

Table A.1

Tier II + Global Hawk Agreement and Amendments: MDA972-95-3-0013

						$1000s TY			
Amend-ment Number	Effective date	Cumulative funding	Total expected value of Agreement	Con-tractor cost share	Phase I	Estimated total cost CLIN 0002+	Total Phase II CPIF Target Price	Phase II CPFF for original SOW	Total Phase II CPFF
Original[a]	5-Oct-94	500	4000		4000				
1[b]	23-Nov-94	4000	4000		4000				
2[c]	6-Apr-95	39000	157988			157988	157348	640.315	640.315
3	29-Sep-95	39319.854	157988			157988	157348	640.315	640.315
4	28-Nov-95	43319.854	157988			157988	157348	640.315	640.315
5	22-Dec-95	44149.133	157988			157988	157348	640.315	640.315
6	25-Jan-96	78149.133	157988			157988	157348	640.315	640.315
7	25-Jan-96	83149.133	157988			157988	157348	640.315	640.315
8[d]	29-Mar-96	86749.133	157988			157988	157348	640	640
9[e]	29-Mar-96	86749.133	157988			157988	157348	640	640
10	31-Mar-96	127749.133	157988			157988	157348	640.315	640.315
11[f]	21-Jun-96	128749.133	158988			157988	157348	640.315	996.479
12	15-Jul-96	128749.133	158988			157988	157348	640.315	996.479
13[g]	23-Aug-96	132937.99	159177			157988	157348	640.315	1185.336
14[h]	29-Aug-96	132887.315	163370			158632	157348	1284.151	5378.497
15[i]	30-Sep-96	132924.623	163408			158632	157348	1284.151	5415.805
16	1-Nov-96	152924.623	163408			158632	157348	1284.151	5415.805
17[j]	26-Nov-96	152924.623	163408			158632	157348	1284.151	5415.805
18	20-Dec-96	169774.623	163408			158632	157348	1284.151	5415.805
19[k]	5-Feb-97	169774.623	163408			158632	157348	1284.151	5415.805
20	25-Feb-97	169775.623	163408			158632	157348	1284.151	5415.805
21	31-Mar-97	177774.623	163408			158632	157348	1284.151	5415.805
22	19-Jun-97	181774.623	163408			158632	157348	1284.151	5415.805
23	16-Jul-97	189774.623	163408			158632	157348	1284.151	5415.805
24[l]	4-Aug-97	200773.623	243785	3100		228010	226726	1284.151	5415.805
25	11-Aug-97	213723.623	243785	3100		228010	226726	1284.151	5415.805
26[m]	1-Oct-97	214473.623	245249	3100		228010	226726	1284.151	6880.287
27[n]	26-Sep-97	214473.623	245249	3100		228010	226726	1284.151	6880.287
28	25-Nov-97	214473.623	245249	3100		228010	226726	1284.151	6880.287
29[o]	4-Dec-97	250223.623	256308	3100		228010	226726	1284.151	6880.287
30[p]	16-Jan-98	250223.623	256308	3100		228010	226726	1284.151	6880.287
31	23-Jan-98	250223.623	256308	3100		228010	226726	1284.151	6880.287
32	10-Feb-98	250938.105	256308	3100		228010	226726	1284.151	6540.437
33	9-Mar-98	250938.105	256308	3100		228010	226726	1284.151	6540.437
34[q]	16-Mar-98	265938.105	271308	3100		228010	226726	1284.151	6540.437
35[r]	31-Mar-98	292476.287	345270	3100		228010	226726	1284.151	112572.402
36[s]	27-Apr-98	292676.287	345270	3100		228010	226726	1284.151	112572.402
37	30-Jun-98	293156.567	345750	3100		228010	226726	1284.151	112572.402
38[t]	30-Jun-98	293156.567	346168	3100		228010	226726	1284.151	112572.402

NOTES: Cumulative funding is the sum of the incremental obligation authority. It represents the maximum liability of the government at any time. Total expected value of the Agreement, by calculation, includes all CLINs plus unallocated CPFF, and the $5.1-million award-fee pool for CLINs 0007 and 0010. Estimated total cost for the original SOW, assumed to be labeled CLIN 0002. Includes both CPIF amount and relevant CPFF when identified. Total Phase II CPIF target price is the Phase II CPIF amount listed on the first page of the Amendment. It includes contractor cost share. Phase II CPFF is the value of CLIN 0005 $3,549,325 + value of total CPFF to date, $1,185,336, which equals $4,734,661. This is $643,836 short of the total value of the CPFF shown on the amendment. That unallocated difference is added to the original CPFF amount. This discrepancy occurs in Amendment 0014. Total Phase II CPFF as listed in the amendments. It includes incremental funds added through additional CLINs as well as unallocated costs. Defined as CLINs 0002—0008, 0010, and 0011.

Table A.1

(continued)

	$1000s TY									
Phase II CLIN 0003	Phase II CLIN 0004	Phase II CLIN 0005	Phase II CLIN 0006	Phase II CLIN 0007	Phase II CLIN 0008	Phase II CLIN 0009	Phase II CLIN 0010	Phase II CLIN 0011	Phase II CLIN 0012	Phase II CLIN 0013
1000										
1000										
1000	188.857									
1000	188.857	3549.325								
1000	188.857	3549.325	37.308							
1000	188.857	3549.325	37.308							
1000	188.857	3549.325	37.308							
1000	188.857	3549.325	37.308							
1000	188.857	3549.325	37.308							
1000	188.857	3549.325	37.308							
1000	188.857	3549.325	37.308							
1000	188.857	3549.325	37.308							
1000	188.857	3549.325	37.308	10999						
1000	188.857	3549.325	37.308	10999						
1000	188.857	3549.325	37.308	10999	1464.482					
1000	188.857	3549.325	37.308	10999	1464.482					
1000	188.857	3549.325	37.308	10999	1464.482					
1000	188.857	3549.325	37.308	22057.818	1464.482					
1000	188.857	3549.325	37.308	22057.818	1464.482					
1000	188.857	3549.325	37.308	22057.818	1464.482					
1000	188.857	3549.325	37.308	22057.818	1464.482					
1000	188.857	3549.325	37.308	22057.818	1464.482					
1000	188.857	3549.325	37.308	37047.818	1464.482	10				
1000	188.857	3549.325	37.308	82686.63	1464.482	10	17703.697	5301.788		
1000	188.857	3549.325	37.308	82686.63	1464.482	110	17703.697	5301.788	100	
1000	188.857	3549.325	37.308	82686.63	1464.482	110	17703.697	5301.788	580.28	
1000	188.857	3549.325	37.308	82686.63	1464.482	110	17703.697	5301.788	580.28	417.914

[a]Original Agreement: 5 Oct 94 is the effective date. The Agreement was actually signed 4 Nov 94. This Agreement covers Phase I of the Tier II+ program.

[b]Amendment 0001: Provides $3,500,000 in incremental funds as planned in initial Agreement, bringing total cost of Phase I Agreement to $4,000,000.

[c]Amendment 0002: This is the initial Phase II Agreement and represents CLIN 0001 and the original SOW. It includes a $157,348,000 total CPIF target price and a $640,315 total CPFF estimated cost and fixed fee. SOW includes design, development, and test of two air vehicles and one ground segment, plus a support segment.

[d]Amendment 0008: Adds task to demonstrate feasibility of AGILE support concepts in ACTD. This is $3.6M total (cost plus NTE fixed fee). Undefinitized in this amendment.

[e]Amendment 0009: Revises Article XXI, System Specification, to conform to the intent under which parties have been operating.

[f]Amendment 0011: Adds CLIN 0003 for Contractor-Acquired Property. This is an NTE effort, capped at $1M, of which $356,164 is approved in this amendment.

[g]Amendment 0013: Adds CLIN 0004, foliage-penetration radar. Also revised the data-rights clause.

[h]Amendment 0014: Adds CLIN 0005 to definitize AGILE support feasability demonstration.

[i]Amendment 0015: Adds CLIN 0006 SIGINT concept-development support. Mislabeled in this amendment; corrected in Amendment 0016.

[j]Amendment 0017: Authorizes $21,400 under CLIN 0003 NTE.

[k]Amendment 0019: Authorizes $232,869 under CLIN 0003 NTE.

[l]Amendment 0024: Complete restructure of program, including cost share and performance fee; redefines agreement completion; provides authorization for fabrication of air vehicles 3 and 4; long-lead for air vehicle 5 (CLIN 0007); adds system specifications to agreement (as a baseline). Cost share is 30/70 TRA/GOV above $206,253,333 until $228,000,000. Additionally, TRA and team agree to spend $3.1 million on SIL development above the contractor portion of the cost share amount. The $226-million value shown is the sum of the estimated cost ($206M), the government's maximum cost share ($15.2M), and the maximum performance fee ($5.2M). The difference between this value and the $230 million shown on the Amendment is $3.5 million, which happens to equal the cost-incentive fee value in Article XIX of the Agreement, amended here. It appears that the total possible fee that can be earned is $8.7 million ($5.2M + $3.5M), but the amendment caps the maximum fee at $5.2 million and states that reduction in fee can be used to offset contractor cost share.

[m]Amendment 0026: Adds new CPFF CLIN 0008 for AGILE support; extends CLIN 0005 at no cost.

[n]Amendment 0027: Extends period of performance for Phase IIB from end date of 26 Sep 97 to 31 Dec 97.

[o]Amendment 0029: Increases NTE for CLIN 0007 (AV3–05) and extends period of performance for Phase IIB through 31 Jan 98.

[p]Amendment 0030: Adds new task to CLIN 0007 for Common Airborne Test Equipment.

[q]Amendment 0034: Increases NTE for Phase IIB (CLIN 0007); adds CLIN 0009 for Phase III planning; authorizes additional NRE task for CLIN 0007.

[r]Amendment 0035: Definitizes CLINs 0007, 0010, 0011 for AV3–5; authorizes CAP under CLIN 0003, revises and adds articles as appropriate. Includes new line on cover page "total agreement value" with value at $347,940,132. CLIN 0010 is integrated sensor suite; CLIN 0011 is ILS tasks. Also defines a total award-fee pool for CLIN 0007 and 0010 of $5,117,730.

[s]Amendment 0036: Adds CLIN 0012 Airborne Communications Node Support, increases CLIN 0009 Phase III planning. Total Agreement Value is $348,140,132.

[t]Amendment 0038: Adds CLIN 0013 in order to recognize the costs associated with the government-directed one-week delay in first flight. Total Agreement Value is $349,038,326.